RESOLVING THE CLIMATE CRISIS

This book brings together a team of renowned social scientists to ask not why climate change is happening, but how we might learn from its human dimensions to raise public and political will to fight against the climate crisis.

Despite efforts for mitigation, global emission levels continue to increase annually and the world's wealthiest nations, including all of the G20 countries, have failed to meet their Paris Climate Goals. In the absence of political will, many have called for individuals to act on climate change by mitigating their own carbon footprint through having fewer children, driving less, using LED lightbulbs, or by becoming vegetarians. While compelling, individual lifestyle changes on this scale are unlikely to prevent climate disaster. *Resolving the Climate Crisis* presents informed solutions for social change that center human behavior and emotions, political systems, and societal structures. Across a series of concise and accessible chapters, authors explore potential solutions to climate change, addressing topics including Indigenous ecologies, LGBTQ+ community engagement, renewable energy technologies, and climate justice. Their expert engagement with the social and behavioural sciences makes this book not only an essential handbook of climate change solutions but also an innovative model for public-facing social science scholarship.

Resolving the Climate Crisis will be an essential resource for students and researchers of climate change, as well as policy makers working to develop meaningful strategies for combatting the climate crisis.

Kristin Haltinner is a Professor of Sociology at the University of Idaho.

Dilshani Sarathchandra is an Associate Professor of Sociology at the University of Idaho.

Routledge Advances in Climate Change Research

Climate Change Action and The Responsibility to Protect
A Common Cause
Ben L. Parr

Climate Security
The Role of Knowledge and Scientific Information in the Making of a Nexus
Matti Goldberg

COVID-19 and Climate Change in BRICS Nations
Beyond the Paris Agreement and Agenda
Edited by Ndivhuho Tshikovhi, Andréa Santos, Xiaolong Zou, Fulufhelo Netswera, Irina Zotona Yarygina and Sriram Divi

Social Transformation for Climate Change
A New Framework for Democracy
Nicholas Low

Exploring Climate Change Related Systems and Scenarios
Preconditions for Effective Global Responses
Jeremy Webb

Resolving the Climate Crisis
US Social Scientists Speak Out
Edited by Kristin Haltinner and Dilshani Sarathchandra

For more information about this series, please visit: www.routledge.com/Routledge-Advances-in-Climate-Change-Research/book-series/RACCR

RESOLVING THE CLIMATE CRISIS

US Social Scientists Speak Out

Edited by Kristin Haltinner and Dilshani Sarathchandra

Routledge
Taylor & Francis Group

LONDON AND NEW YORK

earthscan
from Routledge

Designed cover image: getty images

First published 2024
by Routledge
4 Park Square, Milton Park, Abingdon, Oxon OX14 4RN

and by Routledge
605 Third Avenue, New York, NY 10158

Routledge is an imprint of the Taylor & Francis Group, an informa business

British Library Cataloguing-in-Publication Data
A catalogue record for this book is available from the British Library

ISBN: 978-1-032-56657-3 (hbk)
ISBN: 978-1-032-56739-6 (pbk)
ISBN: 978-1-003-43734-5 (ebk)

DOI: 10.4324/9781003437345

Typeset in Times New Roman
by Apex CoVantage, LLC

CONTENTS

ABBREVIATIONS

ACT	Acceptance and Commitment Therapy
AUM	Animal Unit Month
BP	British Petroleum Corporation
CO_2	Carbon Dioxide
COP	Conference of the Parties
DAPL	Dakota Access Pipeline
EIA	Energy Information Administration
EPA	Environmental Protection Agency
ESA	Endangered Species Act
EU	European Union
EV	Electric Vehicles
GW	Gigawatts
HUG	Hilltop Urban Garden
IEA	International Energy Agency
INTERPOL	International Criminal Police Organization
IOM	International Organization for Migration
IPCC	Intergovernmental Panel on Climate Change
IRENA	International Renewable Energy Agency
KVL-SAR	Kivalina Volunteer Search and Rescue
LEK	Local Environmental Knowledge
LGBTQ+	Lesbian, Gay, Bisexual, Transgender, and Queer
MW	Megawatts
NEPA	National Environmental Policy Act
NGO	Nongovernmental Organization
NIMBY	Not In My Back Yard
NOAA	National Oceanic and Atmospheric Administration

SAI	Stratospheric Aerosol Injection
SRM	Solar Radiation Management
TEK	Traditional Ecological Knowledge
TWS	The Wilderness Society
UN	United Nations
UNHCR	United Nations High Commissioner for Refugees
USDA	US Department of Agriculture
VBN	Values-Beliefs-Norms Framework

CONTRIBUTORS

Ryan Alaniz is a Professor of Sociology at California Polytechnic State University.

Javiera Barandiarán is an Associate Professor of Global Studies at University of California, Santa Barbara.

Nicolas T. Bergmann is a postdoctoral researcher at Washington State University and the University of Idaho.

David Bidwell is a Sociologist and Associate Professor of Marine Affairs at the University of Rhode Island.

Hannah Block is an undergraduate student in sociology at California Polytechnic State University.

Melanie M. Bowers is an Assistant Professor of Political Science at Western Washington University.

Jack DeWaard is the Scientific Director of Social and Behavioral Science Research at the Population Council.

Christina Ergas is an Assistant Professor of Sociology at the University of Tennessee.

Nicole Fox is an Assistant Professor of Sociology at the University of New Hampshire.

P. Joshua Griffin is an Assistant Professor in the Department of American Indian Studies and the School of Marine and Environmental Affairs at the University of Washington.

Kristin Haltinner is a Professor of Sociology at the University of Idaho.

Anne Kristine Haugestad is a PhD candidate at the Norwegian University of Science and Technology.

Shannon Howley is a PhD candidate in the Department of Marine Affairs at the University of Rhode Island.

Aiyana James is the Climate Resilience Coordinator for the Coeur d'Alene Tribe.

Taylor June is a PhD candidate in the Department of Sociology at Ohio State University.

Laura Laumatia is the Environmental Programs Manager for the Coeur d'Alene Tribe.

Cailin Lorek is an undergraduate student in sociology at California Polytechnic State University.

Samantha Noll is an Associate Professor of Philosophy at Washington State University.

Kari Marie Norgaard is a Professor of Sociology and Environmental Studies at the University of Oregon.

Hollie Nyseth Nzitatira is an Associate Professor of Sociology at Ohio State University.

David Osborn is an Adjunct Assistant Professor at Portland State University.

Tristan Partridge is a Lecturer in Global Studies at the University of California, Santa Barbara.

David N. Pellow is the Director of the Global Environmental Justice Project and Dehlsen Professor of Environmental Studies at the University of California, Santa Barbara.

Ryanne Pilgeram is a Sociologist and Regenerative Economies Manager at The Wilderness Society.

Jordan Reeves is the Rural Communities Director at The Wilderness Society.

Dilshani Sarathchandra is an Associate Professor of Sociology at the University of Idaho.

Cedric A. L. Taylor is an Associate Professor of Sociology at Central Michigan University and Visiting Associate Professor at the University of Michigan.

Chloe B. Wardropper is an Assistant Professor in Natural Resources and Society at the University of Illinois, Urbana Champaign.

Cameron T. Whitley is an Associate Professor of Sociology at Western Washington University.

ENGAGING SOCIAL SCIENCE KNOWLEDGE TO RESOLVE THE CLIMATE CRISIS

Kristin Haltinner and Dilshani Sarathchandra

In 1912, the magazine *Popular Mechanics* ran a story (picked up by newspapers in New Zealand and Australia) arguing that the release of carbon dioxide through burning coal would lead to a rise in the earth's temperature in the future (Pester 2021; Sadeghi 2021). By the 1950s, climate change had already risen as a significant concern among climate scientists (Pester 2021). On November 5, 1965, President Lyndon B. Johnson was notified by his science advisory committee that pollution was increasing carbon dioxide levels in the atmosphere with dire impacts on human populations (Nuccitelli 2015). By 1974, climate scientists had demonstrated through climate models that carbon dioxide leads to an increase in the earth's temperature (Nuccitelli 2015). Starting in the 1980s, climate scientists had unified in their demand for change in order to prevent catastrophic climate change (Pester 2021).

Now, in the 21st century, scientists have declared climate change a global emergency, with unwavering consensus and increasing urgency reflected in the Intergovernmental Panel on Climate Change (IPCC) reports, as well as frequent, apocalyptic news headlines. For instance, in 2019, *USA Today* ran a headline indicating that the "climate change apocalypse could start by 2050" as they reported on an Australian study that demonstrated a "doomsday scenario for humans" if significant change is not taken to prevent global emissions. For many people, the climate apocalypse is already a present reality. The Isle de Jean Charles Band of the Biloxi-Chitimacha-Choctaw Indians have lost their homelands as flood waters in Louisiana continue to rise (Faheid and Livingstone 2021). Over 2.3 million Somalians continue to suffer the effects of an ongoing, years-long drought (4 years with an absent rainy season) (United Nations 2021).

Though well-intentioned, much of this scientific messaging and the media coverage of it fails to mobilize the public to action. For example, consider fear-based

DOI: 10.4324/9781003437345-1

messages. Though fear is a useful emotion for getting someone to do an action in the short term, or to get someone to act once, it fails to motivate people into sustained action (Haltinner, Ladino, and Sarathchandra 2021). Rather, the complex, scary, and overwhelming nature of the climate crisis poses a unique problem. It is so amorphous and so large in scope that people become paralyzed, they avoid information on the topic or become apathetic, feeling as though they have no power in the situation (Haltinner and Sarathchandra 2018). In some extreme cases, humans instead turn to conspiracy theories to explain the phenomenon, ironically, in a more manageable (and thus comforting) way (Sarathchandra and Haltinner 2021). It is much less scary, for example, to believe that Al Gore invented climate change to make money or to accrue power rather than to envision an apocalyptic future for all.

That is not to say that nothing has been done. Across the globe, nations have committed to reducing their global emissions. However, despite best intentions for mitigation, global emission levels continue to increase annually (with a notable exception in 2020 as a result of COVID-19 shutdowns). The world's wealthiest nations, including all of the G20 countries, have failed to meet their Paris Climate Goals (Milman 2021). In the absence of political will, and in accordance with neoliberal principles, many have called for individuals to act on climate change by mitigating their own carbon footprint (ironically, a measure developed by British Petroleum to shift the onus of climate change onto the consumer), having fewer children, driving less, using LED lightbulbs, or becoming vegetarians. While compelling, individual lifestyle changes on this scale are unlikely to prevent the climate crisis.

From a scientific standpoint, the reasons for the climate crisis and the steps humans need to take to address it have been clear for quite some time. And yet solutions are nearly impossible to implement because of largely nonscientific obstacles. Because these obstacles are rooted in human culture, including our values, emotions, politics, religious beliefs, group affiliations, news consumption practices, textual representations, etc., we must understand and address them using the tools of social sciences.

The solutions to climate change, such as generating the will to switch over to renewable energy systems or to transform our lifestyles, lie squarely in the realm of the "social" rather than the "scientific." As such, this book brings together some of the top political and environmental social scientists and collaborators to ask not *why* climate change is happening but *how* we might learn from its human dynamics to raise the political and public will to fight against the climate crisis. It asks: *how do we mobilize public will and action to mitigate and adapt to climate change?*

To answer this question, we draw on some of the top minds in social sciences. In this anthology, they share what they have learned from studying human behavior over their careers. These scholars explore how our current cultural stories, largely emerging from the Scientific Revolution in the 16th and 17th centuries, have led to a set of shared cultural values emphasizing differences between invented "types" of humans and between people and the more-than-human world. These stories have served to justify practices of domination and exploitation, resulting in severe social inequality and environmental destruction. The key to mobilizing public will to effectively prevent the worst outcomes of the climate crisis, then, is to radically change our shared cultural ethic and to replace it with values of justice, care, love, and collectivity. These value systems are embraced and practiced by those on the margins of society and must be amplified to replace the dominant, toxic cultural stories. The contributors to this volume consider how to amplify new cultural ways of seeing our relationships to each other and to the planet, as well as specific, creative tools for shifting our cultural narratives. They demonstrate what climate organizing looks like when we put these ethics into practice. Together, this anthology reflects our professional and moral responsibility to address the crisis as scholars and practitioners based in the United States, one of the largest contributors to the ensuing climate crisis.

Part 1: Rejecting our toxic cultural stories

In Part 1 of this book, we explore the toxic, dysfunctional narratives that shape our cultural practices of domination, extraction, and exploitation. In the first chapter, Dr. David N. Pellow, Director of the Global Environmental Justice Project and Dehlsen Professor of Environmental Studies at the University of California, Santa Barbara, examines how neoliberalism and capitalist narratives drive us to focus on individual actions to resolve the climate crisis and concurrently promote the myth that technofixes will curb the worst impacts. Instead, he proposes a more radical approach to mobilizing people to action on the climate crisis: truly democratizing society, ensuring social equity and justice, and taking on the fossil fuel industry directly. Pellow uses the example of organizing around the Central Coast Climate Justice Network (C3JN) to prevent oil drilling in Cat Canyon as a model for an environmental justice approach to climate organizing.

In the second chapter, Dr. David Osborn, a faculty member at Portland State University, explores how settler colonialism has imbued western culture with a set of stories that served to cause and perpetuate the climate crisis. The values associated with settler colonialism teach people that we are separate from nature and one another – that the earth is a tool to use rather than something of which we are a part. Osborn uses insights from his conversations with direct action climate activists in the Pacific Northwest to suggest that to engage with the climate crisis, we need to radically shift our cultural stories. We need to become, what he calls, "earthbound," to come again to experience ourselves as the earth, situated, embodied, and interconnected. A part of this project, he contends, is also to reject the social divisions settler colonialism has created and to work for broader social justice for all as part of and beyond the climate movement.

Complementing these critiques of neoliberalism and settler colonialism, Dr. Christina Ergas, assistant professor of sociology at the University of Tennessee, offers a scathing critique of our cultural stories of gender. Here, Ergas demonstrates that the same processes that led to the extractive, exploitative, and dominating practices of settler colonialism and neoliberalism are similarly rooted in manufactured ideas about gender and the gender binary. She suggests that to mobilize for the climate we need to center educating and empowering women in our work while concurrently working to produce a new collective culture – one that emphasizes collective care over exploitation.

Part 2: Recognizing existing use of alternative stories

With this understanding of how toxic and injurious our current cultural stories are to both humans and to the more-than-human world, our scholars then consider ways of amplifying new cultural stories to replace the old.

Doctoral candidate at the Norwegian University of Science and Technology (NTNU) Anne Kristine Haugestad and Dr. Kari Marie Norgaard, professor of sociology and environmental studies at the University of Oregon, offer a potential mechanism for collective climate action rooted in the traditional Norwegian practice of "dugnad," or shared purpose. The tradition of dugnad combined with a clear picture of the job that must be done to save the climate presents an alternative cultural story that has the potential to convert climate frustration into action and enterprise.

In a creative approach to our call for an analysis of how to motivate people to climate action, Dr. Ryan Alaniz, sociologist at California Polytechnic State University, partnered with two undergraduate students, Hannah Block and Cailin Lorek, to examine climate activism among Gen Z. They find that Gen Z is indeed motivated to fight climate change, despite the despair they often feel about the overwhelming problems faced by their generation. In examining the successes of current climate activism, the team suggests ways to mobilize other young people to leverage sufficient people power to make meaningful change.

Reflecting on revolution – globally and in the United States – mobilization for queer justice stands out as a particularly successful social movement. In their chapter, Melanie M. Bowers, political scientist at Western Washington University, and Cameron T. Whitley, sociologist at Western Washington University, introduce the queer climate movement as a model for successful climate organizing. Using an intersectional, community-based approach focused on justice, the queer climate movement follows the prescriptions offered by Pellow to ensure justice and equity in mobilization.

Part 3: Changing the stories

The authors in the first sections of this book speak to the importance of changing cultural stories. They suggest that our faulty but tightly held (and often invisible) adherence to neoliberal ideas that frame climate change as an individual crisis must be changed. Further, they posit, our reliance on the extractive and exploitative dominance created through the stories of settler colonialism must be changed. This section of the book explores some of the ways such cultural stories can be changed.

Dr. Samantha Noll, associate professor of philosophy at Washington State University, uses social scientific knowledge to consider ways to break through motivational blocks on the issue of climate change. She uses the framework of climate change as a "wicked problem" – one that is especially complex and difficult to solve. Recognizing the lack of public and political will to resolve the climate crisis, in part because of the complexity of the crisis, Noll suggests reframing climate change as a social problem and a wicked problem in public discourse. Doing so allows people to reenvision solutions to the crisis in ways that are perceived as manageable and also in line with their ethical values.

Dr. Kristin Haltinner, professor of sociology at the University of Idaho, explores and challenges narratives that promote individual action to solve the climate crisis. Rejecting neoliberal solutions and green consumption, Haltinner turns to the insight of postmodern feminist theories that demonstrate how to mobilize collective action and change cultural stories. Using examples that range from Bill McKibben's 350.org project, to divestment campaigns, to Fridays for the Future and Meatless Mondays, Haltinner examines how individual actions in partnership with others become collective and lead to meaningful change.

Next, Dr. Dilshani Sarathchandra, associate professor of sociology at the University of Idaho, employs social scientific knowledge about risk perception, emotion, identity, and trust, to consider methods to mobilize action to prevent the worst impacts of the climate crisis. She demonstrates ways to capitalize on pro-environmental attitudes, beliefs, feelings, and identities to motivate the public, even those who may not see themselves as environmental activists, to action on the climate crisis.

In his contribution, Dr. Jack DeWaard, Scientific Director of Social and Behavioral Science Research at the Population Council, draws on psychological theories to suggest a collective approach to moving toward climate action. He reflects on the Acceptance and Commitment (ACT) model, which suggests that change requires us to first accept the dire conditions of the climate crisis and then commit to change. DeWaard uses the example of the Chumash Tribe's work in proposing and managing a new marine sanctuary along California's central coast to demonstrate how this approach can succeed.

Part 4: Amplifying stories on the margins

Given the urgency of the climate crisis, we need to expedite the cultural shift required for full engagement and revolution. In Part 4 of this book, scholars offer ideas for creative methods to shift public ways of thinking about the climate crisis. Though not an exhaustive list of methods for amplifying and embracing new cultural ethics, these creative ideas can serve as examples of the ways this essential work can unfold.

In Chapter 11, Aiyana James and Laura Laumatia, climate change professionals with the Coeur d'Alene Tribe in northern Idaho, begin with a reflection on the ways that settler colonialism has been used to justify natural resource extraction and subsequently damage tribal food systems, ceremonies, and land relationship. They further explore how – in ways that are often ignored – these same ideologies co-opted tribal knowledge systems. From this grounding, they posit that it is incumbent on would-be tribal allies in climate activism to recognize both tribal histories and context for effective climate action, and to recognize how contemporary environmental calls for action must be wary of settler replacement of tribal voices with their own agenda and instead amplify tribal knowledge.

Taylor June, doctoral candidate at The Ohio State University, Dr. Hollie Nyseth Nzitatira, associate professor of sociology at The Ohio State University, and Dr. Nicole Fox, assistant professor of sociology at the University of New Hampshire, take a novel approach to reenvisioning our relationship to the climate: criminalizing climate destruction on a global scale through a restorative and transformative justice model that holds people and nations accountable for environmental destruction. This model forces perpetrators to be accountable for their actions, requiring them to repair the damages they cause, and challenges our cultural practices of accepting extractive practices and destruction as natural, normal, and inevitable.

Dr. Cedric A. L. Taylor, sociologist and filmmaker at the University of Michigan and Central Michigan University, offers the tools of documentary film to challenge the status quo and increase awareness of climate change and its effects, while also shifting the toxic cultural stories that perpetuate extractive practices and social inequality. In this chapter, he demonstrates the important role documentary film can serve to highlight "counterstories" as alternatives to dominant or mainstream cultural stories.

Part 5: Organizing through a new ethic

Climate organizing does not need to wait for a substantial cultural shift to employ ethics of care, love, and collectivity. Rather, it is essential to practice these ethics to both boost this cultural change and ensure successful climate activism. This section highlights case studies that successfully employ a new ethic of care in their efforts to show another world is possible.

In Chapter 14, P. Joshua Griffin, assistant professor of anthropology at the University of Washington, examines community efforts to uphold and adapt marine mammal hunting in the context of rapidly declining sea ice in Northwest Alaska. He features the work of the Kivalina Volunteer Search and Rescue (KVL-SAR), an association of hunters and first responders that plays a critical role in community safety, care, and resilience within the 500-person Iñupiaq community of Kivalina. He describes an ongoing collaboration between KVL-SAR and the University of Washington to co-produce knowledge, action, and capacity in support of locally determined climate adaptation priorities. Drawing on Kyle Whyte's concept of "Indigenous ecologies," the chapter demonstrates ways for Indigenous self-determination in creating and enacting adaptative climate solutions.

Dr. David Bidwell, sociologist and associate professor of marine affairs at the University of Rhode Island, and doctoral candidate Shannon Howley use their expertise in the field of "social acceptance" research to offer suggestions for the construction of wind and other renewable energy arrays. This field assesses the factors that contribute to people's support of renewable energy. They consider the specific case of the Block Island Wind Farm in Rhode Island. They demonstrate how essential it is to include local residents in the planning and installation of renewable energy systems. Specifically, they find success in involving residents in informal communication about the proposed arrays, incorporating residents' concerns and values in the design and implementation, and hiring residents to assist in communication processes. The success of this array, in comparison with others that received widespread resistance from residents, demonstrates a possible way forward in increasing support for locally placed renewable energy systems.

Dr. Tristan Partridge, a lecturer in global studies at the University of California, Santa Barbara, and Dr. Javiera Barandiarán, associate professor of global studies at the University of California, Santa Barbara, look to the past, not to remain wedded to tradition but to understand systems of power and place the future in the activism of the present. To better envision what is possible and how it can be done in a wholistic, sustainable, and democratic way, they take readers on a tour of initiatives – from India to the Scottish Isles, from New York to New Orleans to the United Kingdom mainland – where we see examples of energy transitions as "collaborative projects of justice" – processes in which collective action is centered and justice, consent, and reciprocity are embedded. In short, they offer examples of radical energy transitions beyond a linear shift from fossil fuels and toward renewable energy.

Beyond these specific cases of energy transition, Dr. Ryanne Pilgeram and Jordan Reeves in the Rural Communities Program at The Wilderness Society, explore ways to empower communities in efforts at public land conservation. They especially consider communities that rely on public land for economic viability through extractive practices (timber, mining, drilling) and those that hold historically antagonistic ideas toward environmental movements. Using the specific case of Lincoln, Montana, and the work of The Wilderness Society, they explore the possibilities for lasting models to preserve public land by forming relationships that transcend political partisanship.

Dr. Chloe B. Wardropper, assistant professor in Natural Resources and Society at the University of Illinois, Urbana Champaign, and Dr. Nicolas T. Bergmann, postdoctoral researcher at Washington State University and the University of Idaho, further consider environmental efforts among rural residents in the American West, specifically cattle ranchers. Their chapter puts the suggestions of Pellow and other contributors into action and emphasizes the importance of using local knowledge in the development of climate change mitigation and adaptation strategies.

Through all of these chapters, certain themes emerge: the need for democratic processes, rooted in human agency, local knowledge, and social equity to be at the center of mobilizing against the climate crisis. In example after example, from the C3JN, to the Block Island Wind Farm, to Lincoln, Montana, the social scientists in this volume demonstrate that when change involves people on the ground and considers their lived experiences and concerns, and trust is enhanced, the possibilities for mobilization become endless. By putting into practice a new culture of care, we successfully change values and practices – in both our localized efforts and in our broader culture – to collectively prevent the worst outcomes of the climate crisis. To this end, this anthology presents a clear guide, to be used by organizers, policymakers, and others, to mobilize the public will and take meaningful action to solve the crisis at hand.

References

Faheid, Dalia and Katie Livingstone. 2021. "To Flee or to Stay Until the End and Be Swallowed by the Sea." *Inside Climate News*. July 18. Accessed June 4, 2022. https://insideclimate-news.org/news/18072021/to-flee-or-to-stay-until-the-end-and-be-swallowed-by-the-sea/

Haltinner, Kristin, Jennifer Ladino and Dilshani Sarathchandra. 2021. "Feeling Skeptical: Worry, Dread, and Support for Environmental Policy Among Climate Change Skeptics." *Emotion, Space and Society* 39(2021): 100790.

Haltinner, Kristin and Dilshani Sarathchandra. 2018. "Climate Change Skepticism as a Psychological Coping Strategy." *Sociology Compass* 12.6(2018): e12586.

Milman, Oliver. 2021. "Governments Falling Woefully Short of Paris Climate Pledges, Study Finds." *The Guardian*. September 15. Accessed June 4, 2022. www.theguardian.com/science/2021/sep/15/governments-falling-short-paris-climate-pledges-study

Nuccitelli, Dana. 2015. "Scientists Warned the US President About Global Warning 50 Years Ago Today." *The Guardian*. November 5. Accessed June 4, 2022. www.theguardian.com/environment/climate-consensus-97-per-cent/2015/nov/05/scientists-warned-the-president-about-global-warming-50-years-ago-today

Pester, Patrick. 2021. "When Did Scientists First Warn Humanity About Climate Change?" *Live Science*. December 21. Accessed June 4, 2022. www.livescience.com/humans-first-warned-about-climate-change

Sadeghi, Kenzi. 2021. "Fact Check: A 1912 Article About Burning Coal and Climate Change is Authentic." *USA Today*. August 14. Accessed June 4, 2022. www.usatoday.com/story/news/factcheck/2021/08/13/fact-check-yes-1912-article-linked-burning-coal-climate-change/8124455002/

Sarathchandra, Dilshani and Kristin Haltinner. 2021. "How Believing Climate Change is a "Hoax" Shapes Climate Skepticism in the United States." *Environmental Sociology* 7: 225–238.

United Nations. 2021. "Worsening Drought Affects 2.3 Million People in Somalia." November 19. https://news.un.org/en/story/2021/11/1106222

PART 1

Rejecting our toxic cultural stories

1
A COMMUNITY-UNIVERSITY COLLABORATION FOR CLIMATE JUSTICE

David N. Pellow

The problem: The problem this chapter addresses is how to effectively challenge climate change in a time of massive political polarization and continued emphasis on technofixes and green consumerism.

The solution: The solutions proposed and reviewed in this chapter include initiatives that promote social equity, strengthen democratic engagement and institutions, and keep fossil fuels in the ground.

Where this has worked: This approach worked to promote social equity, democracy, and the reduction of fossil fuel emissions in the Cat Canyon case, and we see similar successes in communities that have shut down or prevented the construction of polluting power plants and other large carbon emitters, from Oxnard, California, to Chicago, Illinois.

Climate change is a problem; technofixes and green consumerism are false solutions

I am an environmental social scientist who has been researching, teaching about, and taking action to address the climate crisis for three decades. There is a widespread consensus (among those not denying or promoting doubt about the facts) on at least two matters related to climate change. First, this phenomenon – also known as global anthropogenic climate disruption – has been in evidence for well over a century (Abram et al. 2016; Lewis and Maslin 2015). And second, we have the tools to solve this wicked problem (see Chapter 7) that presents one of the most prominent and terrifying characteristics of our current historical epoch (Dunlap and Brulle 2015; Hawken 2021; Johnson and Wilkinson 2021).

DOI: 10.4324/9781003437345-3

Unfortunately, two additional observations are relevant here. First, although we have the tools, we lack the political will to mobilize effectively against global climate disruption, and second, far too many of the "tools" that have been developed, applied, and/or proposed amount to technofixes that fail to interrogate and challenge the underlying roots of the problem. In other words, we are so politically and ideologically divided on the question of climate disruption that we are unable to act collectively. Further, so much of what passes for action is focused on individual consumer behavior and nontransformative approaches, all but guaranteeing the continued, unabated approach of the climate apocalypse. Following, I consider what some of these largely futile actions look like.

These tendencies even befall highly educated people who care deeply about the environment. For example, a fellow scholar of environmental politics recently gave a presentation at a public gathering with the aim of delivering a message and guidance on how everyday people can address climate change. I attended, intrigued and excited at the prospect of learning new strategies and tactics for tackling this wicked global problem. Unfortunately, the entire presentation focused on individual, consumerist "solutions" such as electrifying and installing solar panels atop one's home and purchasing a heat pump water heater. While each of these actions can indeed significantly reduce an individual household's ecological footprint, they rely on the assumption that financially privileged, upper-middle-class homeowners are the key to solving our climate crisis and that we can simply buy our way out of this predicament, thus failing conspicuously to challenge the political-economic system that produced and reinforces climate disruptions.

New technologies continue to be developed, posing solutions to the climate crisis that do not rely on changing our consumptive patterns. Solar radiation management (SRM) is a term that is used to describe a particular set of technological proposals aimed at responding to the challenge of climate change. The practice would involve using various techniques (through the use of high-altitude aircrafts and balloons to deposit aerosols in the stratosphere) to reflect a small amount of incoming solar radiation back into space before it has a chance to warm the earth's surface. One such practice is known as stratospheric aerosol injection (SAI), which would involve releasing reflective particles into the stratosphere. As with virtually all technofixes for climate change, SRM and SAI are not fixes at all because at best, they might prevent a minimal level of warming from occurring, and at worst, they siphon vital attention and resources away from more fundamental actions that could be aimed at reducing fossil emissions altogether (mitigation) and altering our behavior and the infrastructure of our homes, businesses, and communities to better prepare for the worst effects of climate change (adaptation) (Jinnah and Nicholson 2019).

With respect to social science research on individualist approaches, Andrew Szasz's (2007) book *Shopping Our Way to Safety* takes aim at the problematic trend whereby people pursue protection from environmental risks through "green" consumerism. This practice amounts to participating in what Szasz terms an "inverted quarantine" – wherein we seek environmental sanctuary for ourselves in a bubble of false promises from corporations selling products that are not nearly as safe as they claim, thus undercutting collective social change efforts that might yield more substantive results. This trend is concerning not only because it reveals how people are supporting "solutions" that fall far short of achieving climate sustainability but also because green consumers may develop a false sense of safety and feel less inclined to participate in collective efforts to support stronger regulation and controls on dangerous industrial activities.

In a related publication, Szasz (2016) extends his earlier critique of consumers to raise important questions about the fixation by *social scientists* on consumption. He rightly points out that consumption has been framed both as a problem (e.g., overconsumption or consumption of foods and products that are harmful to environmental and public health) and as a solution (e.g., green and/or socially responsible consumption). But he argues that scholars must delve deeper into what are the most important driving forces behind our socioecological crises, and consumption is only one of them and is generally less important than others. For example, he points to the important fact of the cumulative effects of per capita consumption on a planet of eight billion people. He considers the role of social inequality as a driver of socioecological harm, which much of the literature on consumption pays less attention to despite research demonstrating that, in many ways, social inequality may be the greatest ecological threat facing us. Furthermore, Szasz notes that consumption is but a single phase of a cycle that includes resource extraction, production, distribution, marketing, consumption, and finally, disposal. He argues that each phase of that cycle produces considerable environmental degradation, particularly those activities associated with mining, manufacturing, and dumping of spent products. In other words, the scholarly focus on consumption often ignores the role of population growth, social inequality, and the majority of the life cycles of various products that thread through our global economies, and that leaves open a large gap in analysis and in the possibilities for developing substantive and effective policymaking.

In the remainder of the paper, Szasz draws from the classic work of Thorstein Veblen, Herbert Marcuse, and others to offer the conclusion that scholarship and policymaking aimed at changing *individual*-level consumption patterns will be extremely limited and even counterproductive if they are not paired with efforts to understand and influence *collective* consumption trends. Moreover, he contends that the literature on consumption generally pays little attention to the largely fixed and inflexible character of consumption that the social geography of cities and suburbs produces, particularly for the majority of persons who have less power and choice over the design of their homes and workplaces. This is a critical lesson to which sociologists studying human-environment interactions should pay heed. Finally, Szasz presents persuasive conclusions for what can be done to address these challenges. He writes that it would be most effective if we advocated for policies that encourage, for example, automakers to improve gas mileage, the building of more energy-efficient homes, and the design of smart cities. I concur and would point to Szasz's (1994) groundbreaking work on social movements that underscores how critical such grassroots efforts are to imagining, devising, proposing, and implementing progressive, socioecologically progressive policies.

Environmental sociologist Norah Mackendrick (2018) builds on Szasz's ideas by demonstrating that the "inverted quarantine" can also be thought of as "precautionary consumption" – those consumer behaviors that seek to avoid toxic products that pervade our marketplace in the 21st century. Mackendrick also argues that this practice is highly gendered and classed, since the burden of shopping "safely" is largely placed on women, and those who can participate in such actions tend to be middle- and upper-class persons, effectively shutting out the majority of people from access to allegedly safer goods. Therefore, she concludes that precautionary consumption preserves gender and class hierarchies and does little to advance the goals of environmental justice.

Another consumerist behavioral "solution" that is also a technofix in widespread usage is the electric vehicle (EV), most of which function with lithium-ion batteries. While individually an EV uses less fossil fuels than a conventional automobile, collectively they exacerbate a number of problems, including the devastation of fragile ecosystems and Indigenous lands during the extraction process for materials for battery production (Center for Interdisciplinary Environmental Justice 2019). The production and use of EVs also contribute to a much subtler but no less serious hazard: a false sense that we are actually addressing the climate crisis.

Furthermore, by their very nature, technofixes like SRM/SAI and consumerist actions like EV usage and green consumerism tend to preserve the political-economic status quo, reinforcing existing power relations in a given society. That problem is anathema to any serious effort to address climate disruption because we know that for effective solutions to emerge and take hold, we need societal transformations that embrace equity and justice at all scales. To the extent that climate change is more or less an "ecological" problem, among the many deficiencies of the technofix and consumerist "solutions" is the fact that they are remarkably *non*-ecological approaches. That is, if the nature of ecological problems is that they reflect disturbances and imbalances within and across webs and networks of *relationships*, attempts to solve such a problem by simply introducing a single technology or product seem clearly inadequate. Furthermore, reducing the problem of climate change to a focus on carbon emissions is woefully reductionist, when the causes of climate change are rooted in ideas, values, and practices that ripple through and shape entire social structures. Another way of making this point might be to say that climate change is actually a result of *multiple* contributing factors, so single-issue thinking and actions will always fall short of successfully tackling it. In fact, it might make more sense to transition away from talking about climate change as a singular phenomenon to start talking about climate *changes* in the plural, since these phenomena affect all regions of the globe in unique and different ways, at distinct levels of intensity, and with particular consequences.

The main point of this section has been to demonstrate that there is a prevalence of what many climate justice activists call "false solutions" to the climate crisis, most of them relying on technofixes and consumerist, market-based activities (Indigenous Environmental Network 2021). These approaches are false solutions because they tend to reinforce the power relationships, social structures, and logics of the institutions, values, and actions that produced the problem of climate disruption to begin with. They are also false solutions because they redirect our attention, energy, and resources away from promoting ideas and actions that are proven to more effectively address climate disruptions. We consider some of these more impactful possibilities in the next sections.

Climate change is a problem; equity and democracy are solutions

As noted earlier, we *do* have the tools to actually confront the problem of global climate change. So, what exactly are they?

1) Sustained efforts to improve social equity and justice, particularly for marginalized populations
2) Concerted action to strengthen democratic institutions and practices
3) Dramatic reduction of greenhouse gas emissions by transitioning away from a fossil fuel economy and by keeping fossil fuels in the ground

Note that none of these three tools involves technofixes or green consumerism. All of them require an active citizenry and robust institutions that support collective action. Later, I consider the scholarly evidence that supports the use of these tools.

Social scientists have authored numerous studies that find that social and political inequality are strongly correlated with and contribute to significant levels of environmental degradation (Downey and Strife 2010). For example, economist James Boyce concludes that *the degree of egalitarianism in a given society appears to be one of the most significant predictors of the level of environmental protection* in that society. Specifically, *societies with stronger measures of economic and political equality are more likely to also have higher overall environmental quality* (Boyce 1994; 2008). The reason for this trend is that communities that are more inclusive and equitable are more likely to feature democratic decision-making and power-sharing, which are practices that also tend to support policymaking that is respectful of our shared ecosystems.

Perhaps the main question that this result might raise is: exactly what mechanisms produce these outcomes? Wilkinson and Pickett (2011) find that increasing levels of social inequality in a society contribute to heightened competitive consumption among its denizens, which, in turn, produces major increases in industrial-scale activities that facilitate climate change in particular and environmental harm more broadly. This area of scholarly research is critical for exploring the linkages between social inequality and environmental decline. However, much of it is focused on economic or political indicators of inequality that do not fully consider the myriad ways in which inequality also operates across race, gender, sexuality, and species.

Climate *injustice* is the term used to describe the fact that climate disruptions affect communities, nations, and regions of the globe very differently. Specifically, Indigenous peoples, communities of color, Global South communities, women, lesbian, gay, bisexual, transgender, and queer (LGBTQ+) people, persons with disabilities, and low-income/wealth populations, are hit hardest by the effects of climate change. They also are the same populations that contribute the least to the problem in the first place since they produce lower levels of carbon emissions, on average – that is why it is considered an injustice (US EPA 2021). Accordingly, climate *justice* is the term used to describe the goal of addressing climate change while simultaneously dissolving and alleviating the unequal burdens created by climate change. The climate justice movement and scholarship emerged directly from the movement and scholarship focused on environmental justice.

Environmental justice scholarship has, for several decades, produced empirical evidence of the strong correlation between race, class, immigration status, and exposure or proximity to pollutants. As with climate injustice, environmental justice scholars find that Indigenous communities, communities of color, Global South communities, low-wealth, and other marginalized communities face disproportionate environmental risk and exposure to a range of threats, including polluting manufacturing facilities, waste storage and management operations, coal-fired power plants, air pollution-heavy transportation corridors, and much more. Furthermore, these same communities tend to enjoy far less access to clean water, healthy air quality, and recreational and green spaces, all of which contribute to higher levels of morbidity and mortality. This thesis is supported by thousands of studies but also features scholarship that examines grassroots political responses to environmental injustice – the environmental justice and climate justice movements (Bullard 2000; Coolsaet 2021; Davies and Mah 2020; Perkins 2022; Pulido 2017).

Thus, the literatures on climate and environmental justice reveals that women, communities of color, low-wealth communities, and Indigenous communities face the greatest threats from climate change but contribute the least to the problem (Ciplet, Roberts and Khan 2015). Going further, however, with respect to the drivers of this crisis, we find that the contemporary "wicked problem" of global climate disruption was initiated by the Industrial Revolution that included a series of European and Euro-American efforts to colonize Indigenous peoples and lands, and the enslavement and forced labor of vast swaths of people across the Global South, which also ushered in the Anthropocene (Estes 2019; Heynen 2016; Whyte 2016). Today we also can observe that there are strong associations between gender inequalities and climate change. For example, Christina Ergas (see Chapter 3) and Richard York (2012) find that in nations where women enjoy access to more equitable political status, carbon emissions are lower. Therefore, initiatives aimed at improving gender equality and gender justice will likely be more impactful if they are paired with campaigns to address climate change and ecological harm. Ergas and York also find that nations with higher levels of military spending also have greater carbon emissions than other nations, reinforcing the decades of scholarship by ecofeminist scholars who have articulated how masculinist policymaking and ecological harms are linked (Gaard 2017).

Human rights advocate Kevin Bales (2016) examined the connections between present-day practices of human enslavement and ecocide and found that many contemporary forms of enslavement are directly implicated in some of the most ecologically destructive industrial activities. These include but are not limited to the extraction and consumption of shrimp, fish, gold, diamonds, steel, cow products, sugar, cocoa, and sandstone. The reason why enslavement is so ecologically destructive is because it frequently involves brutal, industrial-scale extraction of some of the most sensitive ecosystems and vulnerable species, producing unimaginable harm to enslaved workers and more than human populations and spaces.

Finally, while the environmental justice and climate justice scholars have done an excellent job of exploring the significant spatial relationships between marginalized populations and environmental harm, the driving forces behind this phenomenon require more in-depth consideration. Voyles's (2015) concept of "wastelanding" is extremely useful in that regard because it is a racial and spatial signifier that labels particular human communities and landscapes as pollutable and expendable. Historian Marco Armiero's (2021) concept of the "Wasteocene" similarly focuses on the fact that the widespread extractive, violent, wasting relationships that characterize this historical epoch reflect ideological orientations that value certain groups of people and discount and devalue others. The lessons from this literature are clear and profound: ideologies, policies, and practices that produce and enable social inequalities have negative consequences for the people they target, and those inequities drive climate change and ecological unsustainability.

The literature on environmental and climate justice in particular and environmental social science and science more generally suggest quite strongly that we have sufficient knowledge of what is driving our ecological crises. Therefore, various communities and societies could address these challenges by pursuing action intended to strengthen democratic institutions, ensure social equity, and mobilize support for a transition away from fossil fuel–reliant cultures and economies. In the next section, a brief case is considered that illustrates these key points.

Addressing climate change through a community-university collaboration

During the 2018–2019 academic year, I participated in a Sawyer Seminar at the University of California, Santa Barbara, an initiative supported by the Mellon Foundation that brought together faculty and students to focus on climate and environmental justice. We organized public events, conferences, workshops, visiting speakers, film screenings, and courses around two key questions: How are front-line communities disproportionately affected by energy regimes? And how can scholars work with communities to address social and ecological inequities? Thus, the seminar was envisioned as an opportunity for collaboration among university scholars and students, artists, and community leaders. Our core conviction was that grassroots movements and frontline communities are critical to understanding the roots of the problems with carbon-based energy and for devising solutions. My primary involvement in this initiative included leading a class in the spring quarter of 2019, organized around the theme of Participation and Collaboration. That course enrolled both graduate and undergraduate students, who worked with me to design the syllabus and build a lasting collaborative relationship with the newly formed Central Coast Climate Justice Network (C3JN). C3JN was launched as a regional effort to address the climate crisis with the explicit aim of centering immigrants, people of color, Indigenous peoples, and other marginalized communities. In fact, C3JN's cofounders were quite clear that they understood that the key to addressing the climate change crisis was through building multiracial alliances that promote social equity, justice, and democracy. The course was designed so that students were reading published works by scholars and activists on environmental and climate justice policy and movements while working directly to advance knowledge and action with C3JN community leaders on several campaigns.

One of those campaigns was focused on studying and stopping a proposal by three oil corporations to drill 760 oil wells in northern Santa Barbara County's Cat Canyon. Students in my class conducted research on the environmental impact report and found that the companies significantly underestimated the greenhouse gas emissions and volume of water associated with the project. In fact, according to the students' estimates, the combined activities of the three companies involved would have contributed 760,000 metric tons of carbon dioxide equivalent per year to the atmosphere (roughly equal to adding 6.3 million cars to our roadways each year) and would require the use of more than 22 million gallons of freshwater over just 6 years (for activities like well drilling, well cleanout, dust abatement, habitat restoration, and plant relocation, etc.). Students also found that the drilling associated with these projects would threaten to contaminate the Santa Maria River Valley Groundwater Basin, which is the drinking water source for 12 cities and towns with a combined 200,000 residents. They also concluded that the Cat Canyon projects would triple the volume of oil production in the county, a significant increase considering the fact that Santa Barbara County's oil industry is quite large already. Students presented their research to our C3JN partners, assisted with mobilizing other students and community members to attend and offer critical testimonies at numerous public hearings, and offered other forms of key support for promoting climate justice in the region.

The students also met with members of labor unions who were in favor of the project because of the economic benefits they anticipated. And although they eventually agreed to disagree on the merits of the oil drilling proposal, the labor union members responded positively to the student outreach and conversations, noting that no environmentalists or university community members had ever bothered to do so in the past. As each public hearing and protest action occurred, C3JN and the students secured more support from the community, and two of the oil corporations pulled out. Finally, in November 2020, the last of the three corporations withdrew its proposal to drill oil in Cat Canyon (Hodgson 2020). This was a major victory for environmental and climate justice, and the activist community was thrilled. I was delighted to see this community-university collaboration succeed at producing critical knowledge about our local and regional environment that was employed in the service of advancing democratic practices and policies in support of environmental and climate justice. This outcome ensured that oil would stay in the ground – a critical and necessary element in any effort to reduce greenhouse gas emissions and support a transition away from fossil fuels. The process that led to this outcome was also deeply democratic and equitable in that it centered around consensus-based collaborations among a demographically diverse group of students who, in turn, engaged in consensus-based collaborative processes with the C3JN, which is an alliance of organizations whose primary goal is to address the climate crisis and racial justice crisis simultaneously.

Conclusion

Heeding the wisdom of social scientists who have studied climate and environmental justice movements and politics, the community-university collaboration that I co-led focused not on technofixes or green consumerism but rather on mobilizing knowledge and bodies in support of policies and regulatory action that would have direct impacts on climate change in the region. Specifically, by challenging the environmental impact reports that the companies produced and by conducting research on the actual consequences of oil extraction, students and C3JN were able to change the narrative and garner public support for preventing the oil drilling from going forward. This process also involved strengthening democratic practices and social equity because it featured a multiracial group of students collaborating with a multiracial, people of color-led community-based coalition to advance the goals of frontline/fenceline populations who face the brunt of climate change. Community-university collaborations can be effective tools for creating, sharing, and implementing social scientific knowledge in the service of high-impact initiatives to address climate changes and the need for climate justice.

References

Abram, Nerilie J., Helen V. McGregor, Jessica E. Tierney, Michael N. Evans, Nicholas P. McKay, Darrel S. Kaufman, and the PAGES 2k Consortium. 2016. "Early Onset of Industrial-Era Warming Across the Oceans and Continents." *Nature* 536: 411–418. August.
Armiero, M. 2021. "The Case for the Wasteocene." *Environmental History* 26: 425–430.
Bales, Kevin. 2016. *Blood and Earth: Modern Day Slavery, Ecocide, and the Secret to Saving the World*. New York: Spiegel and Grau.

Boyce, James K. 1994. "Inequality as a Cause of Environmental Degradation." *Ecological Economics* 11(December): 169–178.

Boyce, James K. 2008. "Is Inequality Bad for the Environment?" *Research in Social Problems and Public Policy* 15: 267–288.

Bullard, Robert. 2000. *Dumping in Dixie: Race, Class, and Environmental Quality*. Boulder: Westview Press. Third edition.

Center for Interdisciplinary Environmental Justice. 2019. *No Comemos Baterías*: Solidarity Science Against False Climate Change Solutions. *Science for the People* 22(1). https://magazine.scienceforthepeople.org/vol22-1/agua-es-vida-solidarity-science-against-false-climate-change-solutions/

Ciplet, David, J. Timmons Roberts and Mizan R. Khan. 2015. *Power in a Warming World: The New Global Politics of Climate Change and the Remaking of Environmental Inequality*. Cambridge: The MIT Press.

Coolsaet, Brendan. (Ed.). 2021. *Environmental Justice: Key Issues*. New York: Routledge.

Davies, Thom and Alice Mah. (Eds.) 2020. *Toxic Truths: Environmental Justice and Citizen Science in a Post-Truth Age*. Manchester, UK: Manchester University Press.

Downey, Liam and Susan Strife. 2010. "Inequality, Democracy, and Environment." *Organization & Environment* 23(2): 155–188.

Dunlap, Riley and Robert J. Brulle. (Eds.). 2015. *Climate Change and Society: Sociological Perspectives*. Oxford, UK: Oxford University Press.

Ergas, Christina and Richard York. 2012. "Women's Status and Carbon Dioxide Emissions: A Quantitative Cross-National Analysis." *Social Science Research* 41(4): 965–976, July.

Estes, Nick. 2019. *Our History is the Future*. New York: Verso.

Gaard, Greta. 2017. *Critical Ecofeminism*. Blue Ridge Summit, PA: Lexington Books.

Hawken, Paul. 2021. *Regeneration: Ending the Climate Crisis in One Generation*. London: Penguin Books.

Heynen, Nik. 2016. "Urban Political Ecology II: The Abolitionist Century." *Progress in Human Geography* 40(6): 839–845.

Hodgson, Mike. 2020. "Application for Third Cat Canyon Oil Project Withdrawn." *Santa Maria Times*. November 5.

Indigenous Environmental Network. 2021. *Hoodwinked in the Hothouse: Resist False Solutions to Climate Change*. Third Edition. Climatefalsesolutions.org

Jinnah, Sikina and Simon Nicholson. 2019. "Introduction to the Symposium on 'Geoengineering: Governing Solar Radiation Management'." *Environmental Politics* 28(3): 385–396.

Johnson, Elizabeth and Katharine K. Wilkinson. (Eds.) 2021. *All We Can Save: Truth, Courage and Solutions for the Climate Crisis*. London: One World.

Lewis, Simon L. and Mark A. Maslin. 2015. "Defining the Anthropocene." *Nature* 519: 171–180.

MacKendrick, Norah. 2018. *Better Safe than Sorry: How Consumers Navigate Exposure to Everyday Toxics*. Oakland, CA: University of California Press.

Perkins, Tracy E. 2022. *Evolution of a Movement: Four Decades of California Environmental Justice Activism*. Oakland, CA: University of California Press.

Pulido, Laura. 2017. "Geographies of Race and Ethnicity II: Environmental Racism, Racial Capitalism and State-Sanctioned Violence." *Progress in Human Geography* 41: 524–530.

Szasz, Andrew. 1994. *Ecopopulism: Toxic Waste and the Movement for Environmental Justice*. Minneapolis: University of Minnesota Press.

Szasz, Andrew. 2007. *Shopping Our Way to Safety: How We Changed from Protecting the Environment to Protecting Ourselves*. Minneapolis: University of Minnesota Press.

Szasz, Andrew. 2016. "Consumption." In *Keywords for Environmental Studies*, edited by Joni Adamson, William Gleason and David N. Pellow, 44–47. New York: New York University Press.

US Environmental Protection Agency. 2021. *Climate Change and Social Vulnerability in the United States: A Focus on Six Impact Sectors*. Washington, DC: US Environmental Protection Agency, September.

Voyles, Traci Brynne. 2015. *Wastelanding: Legacies of Uranium Mining in Navajo Country*. Minneapolis, MN: University of Minnesota Press.

Whyte, Kyle Powys. 2016. "Indigenous Experience, Environmental Justice and Settler Colonialism." In *Nature and Experience: Phenomenology and the Environment*, edited by Brian Bannon, 157–174. Lanham, MD: Rowman and Littlefield.

Wilkinson, Richard and Kate Pickett. 2011. *The Spirit Level: Why Greater Equality Makes Societies Stronger*. New York: Bloomsbury Press.

2

TOWARD EARTHBOUND CLIMATE MOVEMENTS

The importance of understanding ontology and settler colonialism in engaging the climate crisis

David Osborn

The problem: Despite widespread mobilization on climate change, social movements have not engaged the ontological roots of the climate crisis. In an American context, settler colonialism has received far too little direct engagement both by mainstream and radical climate movements. Settler colonialism is an ongoing process in which relations are structured in such a way that climate change, itself part of a broader crisis of life, is created, allowed, and continued.

The solution: Explicit examination and engagement with dominant worldviews and ways of being structuring American (and to a lesser extent global) society are essential to address the climate crisis. Social movements, and particularly climate movements, have a part to play in either constructing or contesting dominant ways of being and exercising agency in the support of existing lifeways and in the creation of new ones. White settler people and White-dominant movements are identified as especially needing transformation and are the focus of the chapter.

Where this has worked: The direct action climate movement in the Pacific Northwest is one place where this type of cultural change is present. Among the mostly White settler study respondents, there was an emergent transformation of worldviews and ways of being. This transformation positively supported engagement in movement work even while lacking explicit discussion and development within the movement.

DOI: 10.4324/9781003437345-4

In the decade following the turn of the millennium, climate-based social movement organizations emerged resulting in a robust "climate movement."[1] This movement has had enormous impact, including preventing new fossil fuel development, raising awareness about the climate crisis, generating support for renewable energy sources, integrating justice frameworks for climate action, and forcing politicians to follow through on climate-oriented policy.

Yet, these gains have come with significant limitations. While climate change may have been mitigated, at least somewhat, by the collective organizing and actions of millions of participants, this decade has seen the effects of the climate crisis begin to touch larger swathes of society and the emissions trajectory points to dangerous outcomes.[2] Additionally, deep inequities persist with regard to the impacts of the climate crisis and access to resources that support mitigation and resiliency.

The focus in this chapter is on how White settler people, an identity I share, and to a lesser extent non-Indigenous people of diverse identities,[3] can and must engage in a transformation of their relations and worldviews to address the climate crisis, sustain their participation, and act as meaningful allies to Indigenous movements mobilizing for their lifeways. Such an orientation is an essential task for those wishing to act in true solidarity. It presents a parallel body of activity for the types of organizing discussed by Aiyana James and Laura Laumatia in Chapter 11 as well as P. Joshua Griffin in Chapter 14.

I draw on the stories and experiences of people involved in, what I label, "the direct action climate movement." Drawing from my empirical research,[4] I share the experiences of people in a number of groups in the Pacific Northwest including the Sunrise Movement, 350PDX, Portland Rising Tide, Extinction Rebellion, Forests for Climate Resilience, and Climate Direct Action. During my research, these groups organized events such as pressuring elected leaders, youth strikes, cargo ship blockades, and tar sands oil pipeline shutdowns, and engaged in many other actions.

I begin by exploring the climate crisis as rooted in settler colonialism before bringing in the stories of climate activists. Finally, I discuss how the idea of direct action, framed as a ritual, is an especially essential form of action for embodying these ideas.

A crisis rooted in settler colonialism

Settler colonialism works to violently construct and consolidate a particular worldview and set of relations, asserting them as universal, and works to eliminate and erase alternatives, especially those within Indigenous communities. In this sense, settler colonialism is an *ontological* project.

Ontology, an aspect of all cultures, refers to the set of beings and relations that comprise an imagined *world* within which we think, act, and live. The resulting *experience* generates a *worldview* and shapes a *way of being*. In this way, ontology operates at a deeper level than ideology, which assumes the field within which conceptual thought identifies, compares, contrasts, and connects. It is also very important to recognize, especially for those of us operating within the ontology of settler colonialism, that other ontologies don't articulate different beliefs, rather they constitute different worlds in which there are different experiences of being.

The ontology structured by settler colonialism helps us understand all of this better. In its worldview and way of being, it fundamentally positions humans as other than, separate, outside of, or alienated from the rest of life, which is inert and denied subjectivity. This cultural worldview, positioned as "true" and universal, gives rise to ways of relating that are extractive, oriented to domination, and emphasize difference and separateness.

Patrick Wolfe (2006:393) emphasizes that settler colonialism is fundamentally oriented to the control of land and the creation of enabling institutions and is a genocidal project targeting Indigenous people. To accomplish these ends, Burow, Brock, and Dove (2018:63–64) point to how settler colonialism "operates through a reworking of not just the physical landscape but also the ontological landscape" and how this "displacement of Native vision by settler vision makes possible the physical displacement."[5] Wolfe also argues that settler colonialism should be construed as a "structure" rather than an event, which orients us to its ongoing presence in structuring social relations even as it moves on from earlier phases (Wolfe 2006:402).

The outcome of this transformation is a worldview and way of relating that asserts itself as hegemonic, even as it is constantly contested, most significantly by Indigenous communities. This settler way of structuring reality presents an alienated and disembodied orientation that arises from the dualisms of separating culture from nature and the body from the mind. This orientation provides an essential and fertile foundation for the rise, and continuation, of both ecological and social problems, including climate change.

In their research, Burow, Brock, and Dove (2018) focus on the implications for Indigenous-led decolonial practices, initiatives, and struggles. Given the power relations inherent in settler colonialism, this is, of course, essential and deserves boundless attention and resources. That said, the transformation of settler ontology for *settlers and non-Indigenous people* is an essential and parallel body of work. Given that settler colonialism is an ongoing "structure" that is manifesting particular social relations, those most responsible for replicating that structure, namely White settler people, must also change. For climate movements, this perspective and its associated literature are of special relevance given that climate movements often engage (sometimes well, but often poorly) with Indigenous communities. My research, which is presented here, focused on spirituality as a way of getting at these issues.

Emergent transformations and their limitations

Almost a decade ago, I stood on the railroad tracks along the Columbia River about halfway to the Pacific Ocean from Portland, Oregon. It was a crisp, beautiful late summer day. The life around us hummed with the generative possibility of warmth and light. The air was also charged with an energy of social possibility. A 30-foot tripod stood erected on the tracks with an individual suspended at the top. No oil trains would service the Global Partners oil terminal that day. This action was part of a successful movement, sometimes dubbed the "thin green line," that helped prevent the construction of many fossil fuel export terminals in the Pacific Northwest over the last 15 years (Cimons 2020). It was also when I realized that those of us gathered needed to be more of a focus of our transformative intentions.

Specifically, there was a moment when I felt the possibility for the potency and intensity of the action, which involved risking injury and arrest for many people, to help change our relations and orientation to life. It felt possible in that moment to come together and attune our attention to the life, both human and more-than-human,[6] that surrounded us and of which *we were a part* and in so doing to allow the potency of the energy of direct action to sweep us into an experience of our interconnection or our interbeing. At that moment, contrary to so many of the daily practices and ways of life of the dominant culture, we were aligned in that our human living was in generative relations with the living earth. Sadly, at that time, the moment passed without being intentionally named and without an invitation for explicit engagement. We were too busy casting our attention externally, preparing for police repression, and untrained by our movement experience to situate ourselves and the more-than-human life that surrounded us as a focal point for transformation.

This action and almost a decade of movement experience led me to pursue questions related to spirituality, transformation, settler colonialism, and the shape of our relations. I sought to understand spirituality in the direct action climate movement, the construction of collective identities, and participation outcomes. I was surprised by the extent to which there were substantive and emergent transformations of worldviews and identities. These emergent shifts were articulated as essential for participants and supportive of their engagement. Paradoxically, they were also identified as almost never explicitly engaged within formal social movement spaces. This paradox points to tensions and dynamics that must be addressed if addressing the climate crisis requires fundamental shifts in relations and worldviews.

Before diving further into the lessons offered from climate activists, it is helpful to unpack the concept of being "earthbound." Latour (2017) develops an idea of an earthbound identity, a name he attributes to current/future beings that inhabit a religious orientation to particularity and "territorialization." The contrast with dominant worldviews and identities is in being "in the middle of relations that they have to compose one by one," which is to say "nature" as an external concept makes no sense and cannot be an authority, relations are only understood from one's positionality and negotiated within them, and everything is situated rather than universal (Latour 2017:180–183). Haraway (2016:55) in *Staying with the Trouble* plays with Latour's concept and articulates a resonant process of transformation in "ongoing multispecies stories and practices of *becoming-with* in times that remain at stake, in precarious times, in which the world is not finished and the sky has not fallen – yet" (italics added). These ideas strike at a shift in ways of being – or a "mutation" as Latour calls it – toward the relational and away from the separateness structured by settler colonialism.[7]

Surprisingly, the majority of activists I spoke with articulated some level of this earthbound orientation in ways that indicate a reconfiguring of ways of being toward the relational. This orientation included a view of kinship with all life (including, at times, so-called inanimate beings), an orientation to interconnection and entangled identities, a biocentric worldview in which humans were not centered, and an allowance of agency to "more-than-human" beings.

Morgan[8] is one of the people I spoke with. They had engaged in social movements for decades and found a way to sustain their engagement even as social and ecological crises deepened. In our discussion, Morgan talked about the separation settler colonialism has brought into their relations as well as their commitment to transformation:

> Human practice over many centuries, and amplified recently, has been a great severing, a great making of as much of a distance as we can, and it's been functional. We've created this divide between the ways of almost all of the rest of life and what animates us. But it's very strange, this thing . . . So the foundation of my spiritual and religious practice is to notice that this has happened, to refuse it, in other words to consciously live out of the ways of reproduction that makes that and that makes us in that, and to live into a different mode of reproduction which is a different form of being alive. And this other form of being alive is much more . . . Different parts of it are interwoven with each other, the distinctions between things are not hard and fast, and therefore there's a cultivation of indeterminacy. Indeterminacy is not something of weakness but of actual amplification and power.

Morgan also discussed the aspect of kinship and agency identified in this earthbound view in discussing how human interactions with the more-than-human is a "practice that we dance together" which results in "a being that we're making together." Another respondent, Ari, shared their view of how they saw "our body [as] a microcosm of the planet." Respondents also discussed classical understandings of biocentrism including through the explicit use of that concept, in which humans are not centered or prioritized at the expense of others. As Jessica put it, this view "radically [expands] the circle of empathy," and she noted how she has "expanded that empathy towards [oceans]" and "rocks and plants."

Undergirding all of this was a felt sense of interconnection. Many respondents spoke about interconnectedness as a core belief. They used framing such as "non-separateness," "interconnectedness," "interdependence," or a "web that binds all things together." Specifically, Matthew shared that "it's not just some intellectual construct . . . I feel like I am part of, or we're all part of, the fabric of things." Cumulatively, these views supported a view, and experience, that agency resides beyond the human. Sarah offered that she sees trees as "beings" with an "extraordinary amount of intelligence" and even "wisdom." All of this richly maps onto the theoretical work shared earlier as well as that of Anna Tsing and her conceptualization of the "multispecies entanglements that make life across the earth," in which agency is assigned as a fundamental property of all life (Tsing et al. 2017:M2).

This emergent earthbound orientation was also cited as important to sustaining the participation of these activists amid the catastrophe of our times. Chris touched on these themes and brought in his experience of being transformed through living with people in active struggle at the intersection of fossil fuel extraction and ongoing colonization in the Global South. He discussed how he thought of this process as a kind of "re-grounding, attunement, and as I like to think of it, [a] remolding of our outlook on ourselves and the world [that] I think is a crucial element when it comes to all of us [giving] all that it takes." Respondents also shared how they felt brought together as people in struggle but also as being in struggle alongside more-than-human species. Both experiences were explicitly identified as spiritual and supportive to their ongoing engagement.

Additionally, the activists I spoke with discussed in various ways how their earthbound orientations helped sustain them through grounding and integrating intense experiences, supporting the letting go of the illusion of control, allowing and processing grief, and supporting a generative relationship to fear. Rachel, after almost two decades of organizing, had transitioned some of her movement involvement to more directly tending to the processing of these experiences. She highlighted the connection between being in relationship to grief and the climate crisis:

> We live in a grief-phobic society. It's something that's pathologized. . . . We don't honor that grieving is a really important process and part of what makes us human and part of what honors the person or the animal or the place or whatever it is that we lost . . . [and that] those are all really important feedback loops within this living system that we're all a part of that tell us something is wrong. If we live in a place that's being decimated by logging or there's a lot of pollution or what have you, our natural response as humans is to feel sadness around that. And if you're expected to just suck it up, go to work, yes, life sucks but what are you going to do, just put on a happy face and get on with it, that grief has to go somewhere. And so it gets stuffed. And a lot of us have this accumulated grief and other emotions that don't have an outlet. And when that happens, our natural response to these crises that we're living in gets blocked. And in a systems sense, that leads to, well, runaway climate change, for example, because we collectively are not having the scale of response that's needed to respond to this threat. And so how do we unblock that feedback loop? We create spaces where people can actually feel things and where it's okay to feel things and creating safe spaces where people can cry and know that you're okay, there's nothing wrong with you for having strong feelings. And the more that we can do that in community, we start to change our culture and break those things down.

And yet, paradoxically, almost all of the activists discussed the extent to which the explicit discussion or cultivation of alternative spiritualities was almost entirely absent from movement spaces. Most, then, further discussed this absence as detrimental to the mobilization and sustainability of themselves and climate movements. In explaining this paradox, respondents discussed the traumatic experiences that many people have had with dominant religious institutions, leftist taboos against spirituality, dynamics related to vulnerability, and the perceived privateness of spirituality.

When combining the insights from activists with the theoretical literature as well as my own experience, I'm struck by the need to explicitly engage questions of spirituality, ontology, and the structure of our relationships in movement work. Among those I spoke with, divergent spiritual orientations and ways of being had clear impacts for movement activists on both the motivation for participating in high-risk activism as well as sustaining that participation. I am surprised and excited by the extent to which alternative worldviews may be germinating quietly beneath the surface within climate movements.

Becoming earthbound

In considering the implications of this research as it relates to climate organizing, a re-grounding in the crisis of the Anthropocene[9] is essential. First, the climate crisis is but an aspect of a broader crisis that is fundamentally about our relationship to life. Solving excessive greenhouse gas emissions, for example, would do very little to address the mass extinction event we are in the midst of. Second, current climate efforts also exacerbate inequality and oppression for humans as they play out across gender, race, and class (to name but a few). This latter point is, of course, a central aspect of the critique that the direct action climate movement made alongside Indigenous and frontline communities of mainstream climate and environmental movements, captured in the concept of climate justice. It is also another aspect of how we relate to life, in this case, human life.

With this view in mind, it becomes essential to seek better ways of organizing. Addressing the climate crisis may be part of a transition away from the dominant cultural identity and worldview of separateness cultivated, slowly and with constant contestation, over thousands of years in Western Europe and then violently exported throughout the world. This identity and worldview are the outcome of settler colonialism as a project that mobilizes people in a particular way and with particular relations with the rest of life. It was and is ontological. Thus, it is essential to recognize the *climate crisis as a cultural problem*. Intentionally mobilizing ourselves as a different kind of people, and thus how we relate to other beings, must be *a part* of what movements are doing.

I am calling this project "becoming *earthbound*," at least for White settler people. It is important to remember that some peoples are already earthbound in their own diverse ways and that the project of becoming earthbound is necessarily *plural* in that our relations are always situated, negotiated within that context, and only understood from our positionality (Latour 2017:180–183). In the movement I studied, I saw this work clearly present though unwilling, or perhaps unable, to come out of the shadows.

Critical here is whether the *people* who social movements are trying to mobilize are aligned with the *people* as constructed by the dominant culture. Finding *alignment* with other peoples and movements with which to make common cause becomes essential with this view, and we can only do this if we are clear on who we are and who we are trying to become.

To help movements and scholars more clearly engage these questions, I generated a framework that centers this idea of "becoming a people."[10] This framework resonates with preexisting movements, particularly those (such as the direct action climate movement) that seek to radically change social relations as they relate to class, sexuality, gender, and race. There is already a strong precedent for trying to become a different kind of being in many movements. Climate organizing must radically change how the dominant culture understands and structures relations with the more-than-human world, or what the West knows as "nature."

To do this, or any transformative work, there is a need for stories and mythologies as well as practices that inhabit those mythologies (such as through rituals and ceremonies). These *stories* contain a set of religious beliefs as well as the language, signs, and symbols that reference these beliefs. The *practices* then actualize and create an experience of inhabiting those beliefs. These are all different words to talk about religious dynamics, which were identified by Morgan as "a word for one of the practices that mobilize people as people, in other words, groups, particular kinds of people." Last, there is the strength of the *connectivity* individuals experience with the stories and practices as well as with the people which one is becoming. This is a different word to talk about spiritual dynamics, identified by the same respondent as "a word to describe the texture and the strength of certain sorts of connection one has through religious practices."

This framework is visualized in Figure 2.1 and illuminates a process that is operational for *all* cultures. The dominant culture is constantly tending to the stories, practices, and felt connectivity that stabilize the dominant way people are mobilized in Western society, which are perhaps most fervently taken up by White settler people. This includes the alienated, dualistic, extractive, and dominating ways of being that undergird the widespread destruction of life currently underway. If the direct action climate movement, and social movements broadly, wishes to cultivate a mobilization of people as something other than atomized individuals in a dualistic and objectified relationship with "nature," they will have to be very intentional as it will go against so many of the stories and practices that are hegemonic within the dominant culture.

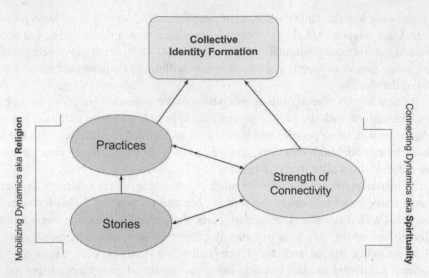

FIGURE 2.1 Theoretical Framework for Collective Identity Formation (Osborn 2023)

Conclusion

My experience on the railroad tracks many years ago still resonates. Through continued listening to the teachings of that experience, I have been speculating on the potential potency of direct action for the work of becoming earthbound. Morgan's conversation with me also identified this possibility:

> I think direct action gives the participants the taste of a different reality . . . One is participating with a different way of practicing life in which what we do as human living beings, like what we literally do, the movements we make with our arms and our mouths, our motion and then our direct participation with others, like immediately around us, becomes world historical. Literally it becomes the place in which, at least in a small way, in a fragmented way, it becomes the place in which change is either happening or not happening. It becomes important. It becomes part of the life of the world.

With this view in mind, direct action can become a ritual in which *the participants* are as much the "target" of the action as the fossil fuel industry, bank, or mining operation. In other words, the power and energy of direct action in even momentarily constructing different relations and a different world can be used to help bring about their stabilization among participants. It is a moment in which we can act out of partial and emergent worldviews and relations, feel them coming alive within the life of the world, and then help breathe them more deeply into our lives.

The potency arises from the tension, risk, and rupture that direct action often creates through its intentional denial of established law, practice, or conventional behavior. Possibilities for transformation also arise through the embodied experience of responding to, or even being in, the crisis of these times. For so many of us, especially White settlers, our day-to-day actions simply do not seem to correspond or relate to the extraordinary violence and destruction of life occurring all around us. On the contrary, our actions are often directly contributing to it. When we sit down and block a train carrying coal, obstruct an exploratory oil vessel on its way to the Arctic, or blockade the entrance of a bank profiteering from the tar sands, we come into a momentary alignment with life. At the same time, for many people and communities this violence and destruction are felt daily. It is thus also essential that this alignment involves explicit solidarity, a practice of following the leadership of those most impacted, and an ongoing commitment of our lives and actions to collective liberation.

Acting in this way, we cultivate an alignment, presence, or *being in* these times that has its own power. Latour (2017) argues that the climate crisis heralds the possibility of a resensitization to place in which individuals' interests arise in relation to the particular realities of the places they consciously and intentionally inhabit. Haraway talks about this as "becoming-with." Tsing emphasizes our "contamination," or inherent transformation through encounter, in our participation in the "multispecies entanglements that make life across the earth" (Tsing 2015:28; Tsing et al. 2017:M2). These effects require our intentional presence to these realities alongside our embodied action. Otherwise, for some, it is all too easy to look away and continue to cultivate stories and practices consistent with the dominant cultural orientation of separation and domination.

With regard to settler-Indigenous dynamics, this work will also allow White settler and non-Indigenous groups to act in greater alignment with Indigenous communities and organizations. Within the frameworks offered here, it is useful to note that Indigenous resistance has often been about defending, recovering, and cultivating mobilizations of being a people that are counter to that of dominant settler culture. As demonstrated at the Standing Rock mobilizations in 2016, these forms of resistance can be extraordinarily attractive to White settlers and non-Indigenous people. In the context of my discussion here, this may be precisely because they present a deeper, broader, and clearer critique of what we are fighting against and fighting for. In other words, the Indigenous people leading the mobilizations at Standing Rock were very clear and explicit that their resistance was part of becoming, or remaining, a different type of people for which, to use one poignant example, water *is* life. However, unless White settler movements more explicitly cultivate their own project of becoming different peoples, they risk participating in the very cultural appropriation and destruction of Indigenous ways of being that has been an essential part of the colonial project.

My conversations point to one aspect of the broader climate movement in which there is an emergence of an attempt to mobilize *different peoples* among the mostly White settler participants. I made the case that this work is of particular importance for White settler people, needs more explicit attention, is essential for addressing the climate crisis, and has the potential to provide motivation, energy, and a sense of wholeness that will fuel movements. It is my view that for both social movements as well as social science researchers, we should become more reflexively aware of how we are constantly engaged in the ontological process of *becoming a people*. This focus also reveals the agency we have in the choice, if consciously realized, to either reify or transform how we are culturally mobilizing ourselves as a people.

Notes

1 For several early examples of movement and scholarly identification of a "climate movement," see Building Bridges Collective 2010; Moore et al. 2011; Tokar et al. 2010.

2 It is essential to note that climate change has long been experienced by Indigenous communities and frontline communities. The language here is meant to indicate that the impact of the crisis has recently become felt broadly across communities and places.

3 Certainly, some of what I engage here is relevant to non-Indigenous peoples of diverse identities. However, it is my view that there are significant differences and particularities of the processes that different communities must transit for this work. Many of these communities – for example, Black and Latinx – already have substantive momentum on projects of ontological transformation connected to their intergenerational liberation struggles. It is most appropriate to point to the relevance of my findings for White settler people and White-dominant aspects of the climate movement.

4 See Osborn 2023.

5 Federici, in her essential work *Caliban and the Witch*, explores the concurrent and enabling transformation of ontology and associated restructuring of relations in Europe. She makes the argument that the violent precapitalist accumulations and enclosures happening in the 16th and 17th centuries, which particularly targeted women through the witch hunts, worked to complete a transformation of European relational ontologies,

> This is how we must read the attack against witchcraft and against that magical view of the world which, despite the efforts of the Church, had continued to prevail on a popular level through the Middle Ages. At the basis of magic was an animistic conception of nature that did not admit to any separation between matter and spirit, and thus imagined the cosmos as a *living organism*, populated by occult forces, where every element was in "sympathetic" relation with the rest. (Federici 2014:141–142)

6 I am using the phrasing, the "more-than-human," coined by David Abram (1997). My intent in doing so is to acknowledge the instability in the Western dualistic concept of "nature" in its strong subject/object dualism.

7 It is important to note that Haraway (2016) clearly points to the ongoing presence of Indigenous cultures – largely through an in-depth case study of the Diné people – as an embodiment of this orientation. Latour's treatment, unfortunately, remains more theoretical. From my perspective, it is essential to notice the multiplicity of ontologies *present* in the world today, especially those of Indigenous communities, and then inquire about their relations to systems of power and problems such as climate change. Then we must ask situated and appropriate questions about disrupting and changing dominant ontologies from the positionality of our identities.

8 The names of all respondents have been changed and given pseudonyms to protect their identity and address security concerns in doing research on social movements that are the focus of state repression. I use the gender identities used by respondents, they/them is here used as a nonbinary gender pronoun.

9 The concept of the Anthropocene has been criticized in many places. I am sympathetic to the critiques articulated by Haraway (2016) and Tsing (2015), among others. I share many of their concerns including the way in which it centers humans rather than particular cultures or ways of being. I use it here with reservations and given its prominence and use as shared language.

10 For a more in-depth exploration, see Osborn (2023) pages 96–98 for a development of the concepts of spirituality and religion and pages 202–205 for the framework.

References

Abram, David. 1997. *The Spell of the Sensuous: Perception and Language in a More-Than-Human World*. New York: Vintage.

Building Bridges Collective. 2010. *Space for Movement? Reflections from Bolivia on Climate Justice, Social Movements and the State*. Leeds: Footpring Workers Co-op.

Burow, Paul, Sandra Brock and Michael R. Dove. 2018. "Unsettling the Land." *Environment and Society* 9(1): 57–74.

Cimons, Marlene. 2020. "The Making of the 'Thin Green Line.'" *Public Broadcasting Service*. Accessed November 14, 2023. https://www.pbs.org/wnet/peril-and-promise/2020/04/the-making-of-the-thin-green-line/

Federici, Silvia. 2014. *Caliban and the Witch: Women, the Body and Primitive Accumulation*. Brooklyn, NY: Autonomedia.

Haraway, Donna J. 2016. *Staying with the Trouble: Making Kin in the Chthulucene*. Durham, NC: Duke University Press Books.

Latour, Bruno. 2017. *Facing Gaia: Eight Lectures on the New Climatic Regime*. Medford, MA: Polity.

Moore, Hilary and Joshua Kahn. 2011. *Organizing Cools the Planet: Tools and Reflections to Navigate the Climate Crisis*. Oakland, CA: PM Press.

Osborn, David. 2023. "Earthbound in the Anthropocene: Spirituality, Collective Identity, and Participation in the Direct Action Climate Movement." PhD Dissertation, Sociology, Portland State University.

Tokar, Brian and Eirik Eiglad. 2010. *Toward Climate Justice: Perspectives on the Climate Crisis and Social Change*. Porsgrunn, Norway: Communalism Press.

Tsing, Anna Lowenhaupt. 2015. *The Mushroom at the End of the World: On the Possibility of Life in Capitalist Ruins*. Princeton: Princeton University Press.

Tsing, Anna Lowenhaupt, Nils Bubandt, Elaine Gan and Heather Anne Swanson. (Eds.). 2017. *Arts of Living on a Damaged Planet: Ghosts and Monsters of the Anthropocene*. Minneapolis: University of Minnesota Press.

Wolfe, Patrick. 2006. "Settler Colonialism and the Elimination of the Native." *Journal of Genocide Research* 8(4): 387–409.

3

GENDER AND CLIMATE JUSTICE

Christina Ergas

The problem: Social inequity in all its forms (e.g., class, caste, race, gender) serves as the basis and justification for the uneven distribution of environmental resources and risks. More environmental goods are distributed to those in dominate positions, while environmental burdens are distributed to those in subordinate positions. The more unequal a society, the more environmental problems exist generally. One way this is expressed is through higher greenhouse gas emissions in nations with lower proportions of women in decision-making bodies, which exacerbate climate change globally.

The solution: Cultural and institutional mechanisms are needed to ensure that everyone has equitable access to education, material resources, land, and decision-making structures. Particularly, communities with educated girls and women who have access to meaningful work and political empowerment tend to have better environmental outcomes, especially related to climate change. Cultural norms and economic incentives should prioritize low-emissions work that contributes to gender justice as well as community and ecosystem well-being.

Where this has worked: The Kenyan Green Belt Movement empowered girls and women to have positive effects on local communities and environments. Through environmental and political education, women learned that local leaders were more concerned with advancing environmentally destructive industrial development than considering the well-being of the

DOI: 10.4324/9781003437345-5

people who lived in the communities devastated by development. This group of women took matters into their own hands to restore their ecosystems by planting millions of trees as well as demanding better political representation. Resulting ecosystem restoration revitalized their communities and local economies and drew down carbon dioxide, which mitigates climate change.

Academic scholarship continually points to the need for social equity to address the climate crisis. In this volume, the urgency of equity-driven, democratic organizing is seen in a number of chapters, including those about community-university collaborations by David N. Pellow (Chapter 1), case studies about public land by Pilgeram and Reeves (Chapter 17) as well as energy systems by Bidwell and Howley (Chapter 15) along with Partridge and Barandiarán (Chapter 16). As these authors have demonstrated, inequalities in income, race, and gender, among others, greatly affect well-being and have environmental implications that cannot be understood in isolation from one another.

Flint, Michigan's lead-contaminated water crisis is a recent example of these dynamics. Gender, race, and class disparities intersect with the legacy of industrial environmental contaminants to harm people and local ecosystems. Flint's residents are primarily racialized as Black, and 45% live below the poverty line. Flint also has a larger share of children living in unmarried woman-headed households (55%) than the rest of Michigan, with married households being predominate at 65%. After General Motors and other industries left, Flint remained an economically depressed town in debt. In 2013, state officials looked for ways to cut costs in Flint. They chose to quit piping treated water from Detroit and instead temporarily pump water from the Flint River. However, the river water was highly corrosive due to previous industrial activity, and officials failed to treat it (Natural Resources Defense Council 2018). As a result, lead, a heavy metal toxic to human health, leached out from aging pipes into residents' homes. Over the course of 18 months, 9,000 children were exposed to lead-contaminated water, and elevated blood-lead levels in children citywide nearly doubled – and nearly tripled in certain neighborhoods. Between 2013 and 2016, little action was taken by city and state officials to correct the situation, and the residents of Flint were forced into lengthy litigation efforts against their local government to ensure action. Today, lead levels in city water stand below federal action levels; however, the cleanup process and pipe upgrades remain slow (Natural Resources Defense Council 2018). This still unfolding crisis is the focus of sociologist and filmmaker Cedric A. L. Taylor's documentary *Nor Any Drop to Drink* which he discusses in Chapter 13 of this volume.

As with the Flint water crisis, a rich body of empirical work demonstrates links between social inequalities and environmental problems. Increased income inequality is associated with biodiversity loss, resource consumption, waste generation, toxic emissions, and water pollution as well as lower survival rates of children under the age of 5 years (Agyeman, Bullard, and Evans 2003). Gender inequality is associated with greater carbon dioxide (CO_2) emissions. Conversely, efforts to introduce sustainability measures are enhanced by women's participation in community decision-making bodies, which lead to better protection of common property resources and lower CO_2 emissions (York and Ergas 2011; Ergas et al. 2021). This research suggests that eradicating myriad social inequalities plays a vital role in realizing climate change mitigation goals. Understanding the relationship between environmental outcomes and different dimensions of inequity allows us to consider forms of human development that simultaneously serve people and the environment.

This chapter focuses on the need to center gender equity in climate change mitigation and adaptation efforts. Relying on insights from ecofeminist scholarship, I build off a recognition of how interlocking systems of domination – such as capitalism, heteropatriarchy, white supremacy, settler colonialism, and human dominionism,[1] among others – lead to both social and environmental ills. In this chapter, the relationship between intersecting inequities related to gender and environmental harms are outlined. I also take the lessons offered by ecofeminism to demonstrate that women and girls need to be central to climate organizing and that this must include a radical shift in our shared ethic away from one of exploitation, extraction, and greed, and toward one focused on mutual care, love, and collectivity.

Social construction of inequity

To understand gender and climate change scholarship, we must first recognize that gender is a social construction. This means that gender is re/created through social interactions and codified into social institutions. Gender is not an essential or biologically determined part of one's identity. Gender also is not a static category in that it differs across contexts and cultures and changes over time. The construction of gender begins for most individuals before birth when they are assigned a gender based on what their genitalia look like in utero. One's sex, then, is formally assigned at birth, and, in US culture and many cultures globally, it is assumed that one's gender aligns with their sex assignment (i.e., a male sex assignment is gendered masculine, and female is feminine). From then on, individuals are socialized to meet the expectations, norms, and roles of their assigned gender (Lorber 1994). While gender might be a constructed category, sociologists have long argued that if people perceive situations as real, "they are real in their consequences" (Thomas and Thomas 1928:572). In the context of climate change, the structural organizing of gender and its associated norms and ideas are reflected in the broader cultural stories around domination and extraction.

Just as gender is socially constructed, so too are human relationships to their environment. Though dominate culture in the United States reflects a settler-colonial ideology centered on extraction and domination (as discussed by David Osborn in Chapter 2), there are alternative models of being. For example, in some Indigenous and subsistence-based societies, cultural understandings of the earth are of a nurturing mother who provides for the needs of the people, and the people see themselves as of the land and part of the ecosystem. Kimmerer (2013), a biologist and member of the Potawatomi Nation from the now Great Plains region of the United States, discusses her cultural and spiritual connection to the natural environment. She describes listening to the wind and trees as a means of learning from her "more-than-human" community. In alluring prose, she advances a culture and economy "aligned with life, not stacked against it," an economy that stands in stark contrast to our current economic system (Kimmerer 2013:377). She personifies capitalism as a monster of overconsumption, whose greed and bottomless hunger cannot be sated, because its emptiness stems from a deep loneliness and longing for mutuality. She describes her culture as one of gratitude and reciprocity, which appreciates the gifts of nature and is responsible for caring for these gifts and giving back.

For Western Europeans, this construction of nature changed around the time of the Scientific Revolution, which began in the 16th century (Merchant 1990). Nature, as feminine, morphed into a wild and chaotic entity in need of controlling – particularly her sexual appetites and biological reproductive capacity. Along with this new construction came new social norms toward the earth, as Europeans went from venerating nature to seeing it as instrumental for their material gain and therefore exploitable.

Causes

Ecofeminists have long examined how, during the Scientific Revolution, elite Western European men employed religion, philosophy, and science to create hierarchical categories of humans that legitimized inequality, exploitation, and violence. They exported these constructions globally through colonization and the expansion of capitalism leading to an increased acceptance (often enforced by coercion and violence) of a gender binary and normalization of extractive practices and human domination over the environment. The newly emerging elites of global capitalism profited from exploiting these inequalities.

Social inequality is a hierarchical relationship wherein resources, such as environmental goods, power, and wealth, are unevenly distributed. Specifically, in gender hierarchies, men in most societies hold more power, resources, status, and wealth than do women and gender nonbinary individuals. This relationship can become more or less uneven due to other social hierarchies related to class, race, ethnicity, nation, and caste, among other factors. Environmental risks and vulnerabilities are unevenly distributed, disproportionately burdening those with less power and status (Gaard 2017; Pellow 2014; Ergas 2021). Specifically, those with wealth and power enjoy goods, property, and clean environments.

However, material wealth for some comes at a cost for others. Specifically, the costs of wealth result in degraded ecosystems, stolen Indigenous peoples' lands and resources, health problems for poor or working-class laborers, and women's unpaid reproductive labor (biological reproduction, unpaid care work, and domestic labor). For example, the Dakota Access Pipeline (DAPL) is a means of transporting crude oil from North Dakota to Illinois. Built by Energy Transfer Partners, the pipeline supplies energy to US consumers, mostly settlers on Indigenous lands. Part of the pipeline runs under the Missouri and Mississippi Rivers and Lake Oahe, and within half a mile of the Standing Rock Sioux Reservation. Aside from the greenhouse gas emissions the pipeline produces, oil spills are common. DAPL leaked at least five times in 2017 (Brown 2018). The pipeline threatens to contaminate the ecosystem and drinking water that sustain the tribe as well as destroy important cultural, spiritual, and burial sites (Harvard Law 2023).

Inequalities are used to justify brutal and violent means of control, such as colonization, enslavement, labor exploitation, dispossession, displacement, genocide, resource exploitation, and the witch trials (Merchant 1990; Gaard 1997). Coercive forms of social control benefit a small group of social elites who reap the material benefits. As in the Standing Rock example, when water protectors stood up to protect Indigenous waters from DAPL, state and government officials sent police in riot gear and the National Guard to protect the company. Police raided encampments, arrested hundreds of people, and used violent means to subdue protestors (Monet 2016). For these reasons, ecofeminist scholars argue that there is a clear connection between social inequalities and environmental problems; they enable each other (Gaard 2017; Pellow 2014).

The central concern of ecofeminism is that through the creation of these contrived hierarchical binaries, the subordinate category of each is relegated to exploitable resource, which discursively and politically legitimates their exploitation. Gender, then, becomes a pattern through which humans engage with nature: gender divisions of labor affect how men and women interact with and experience local environments (Ergas 2014). Women's cultural and political subordination, compounded by their lack of wealth and social capital, makes them particularly vulnerable to environmental risks. Intersecting inequalities make women and children among the most vulnerable to climate-related disasters, particularly in Global South nations. Other factors increase risk, including poverty, rurality, Indigeneity, among others. For example, Bangladeshi women and children comprised 90% of the 140,000 people who died in the 1991 cyclone (Schmuck 2002). Bangladeshi norms of proper dress, such as long saris, limited women's mobility and made them more vulnerable to storm surges (Cannon 2002). In their gendered roles as caretakers, many women also had young children to save. Moreover, many women never learned to swim due to cultural expectations of modesty (Cannon 2002).

Poor rural women in the Global South disproportionately and distinctly experience environmental burdens and resource degradation. Gendered divisions of labor often position rural women as the primary fuelwood, water, and subsistence food gatherers. Resource degradation may force women to travel farther for resources, which can increase their exposure to life-threatening toxins and diseases (Rocheleau, Thomas-Slayter, and Wangari 1996). For example, during the 1970s, rural women in Nepal were able to collect fuelwood in 2 hours, but as forests were cleared throughout the next decade, their collection time increased to an entire day and involved walking through rough terrain (Shandra, Shandra, and London 2008). As water collectors, women face exposure to malaria, which is endemic in many parts of Africa and Central and South America (Denton 2002). Researchers theorize that women's different environmental experiences also affect their environmental concern.

The hierarchical social constructions of gender and nature have profound consequences. This construction has led to an unquestioned acceptance of extractive practices and environmental domination which has resulted in biosphere disruptions, such as climate change, and women's subsequent disproportionate vulnerability to increasing natural disasters. These examples reflect a global trend whereby women experience more vulnerability to the effects of climate change than men (Rocheleau, Thomas-Slayter, and Wangari 1996). Unequal burdens will only worsen as the intensity and frequency of such events increase due to climate change (Ergas, McKinney, and Bell 2021).

Solutions

Just as social inequalities are highly correlated with environmental harms, the converse is true. More equity leads to better environmental outcomes for everyone. Given that disparities in access to resources, education, and decision-making power have negative consequences for people and ecosystems, it stands to reason that eliminating these barriers for marginalized communities is a way toward climate mitigation and adaptation. If those with less power gain equitable access to education, political spaces, and decision-making bodies, researchers have hypothesized that the prevalence of these environmental problems, as well as human health disparities, are likely to lessen. In addition, many argue that our cultural stories need to shift to 1) emphasize environmental and human health and well-being as well as 2) our collective need to care for one another and the planet.

Women's education and leadership

Educating girls generally, and about family planning in particular, combined with making reproductive healthcare more accessible give young women more options and reduce fertility (Lutz and Kc 2011). Reducing population growth indirectly reduces demand for environmental resources and generally improves living conditions (Dietz, Rosa, and York 2007). However, some argue that focusing on the reproduction aspects of educational attainment is limited and threatens to reinstate problematic population control measures in a seemingly more enlightened veneer (Wilson 2017).

Education for women also results in families that are more prepared to make informed decisions about how to prepare for and adapt to climate-related disasters. For example, Kwauk and Braga argue that:

> when girls and women are better educated and included in decision making at all levels, their families and communities are more resilient and adaptable to economic and environmental shocks and are better able to plan for, cope with, and rebound from climate crises. Data suggest that there is a strong positive association between the average amount of schooling a girl receives in her country and her country's score on indexes that measure vulnerability to climate-related disasters.
>
> *(Kwauk and Braga 2017:5)*

In short, climate organizing must rely on providing reproductive health education and access for women, providing opportunities for education more broadly, and empowering women with the knowledge and confidence to participate in the public sphere.

An example of a movement that took a similar strategy is the Kenyan Green Belt Movement, founded by Professor Wangari Maathai in 1977. This movement led by women sought to undo the damage caused by state development. Through environmental and political education, women learned that local leaders were more concerned with advancing environmentally destructive industrial development than the well-being of the people who lived in the communities devastated by development. This group of women took matters into their own hands to restore their ecosystems by planting millions of trees as well as demanding better political representation. Ecosystem restoration revitalized their communities and local economies by reviving the watershed and restoring soil health, which allowed for more agricultural food production and fuelwood for energy. Ecosystem restoration also draws down CO_2, which mitigates climate change (The Green Belt Movement 2023). Investing in girls' education to promote their environmental leadership may prove valuable as research demonstrates that women's participation in decision-making bodies leads to better environmental outcomes.

Having more women in political positions of power or in climate organizing more broadly results in greater levels of environmental policy. For example, the United Nations (2007) reported that between the years 1990 and 2004, 18 of the 70 most developed nations in the world had stabilized or reduced their carbon emissions. Of these 18 nations, 14 had a greater than average percentage of women as elected representatives. Shandra et al. (2008) found that nations with a higher proportion of women's nongovernmental organizations also had lower per capita rates of deforestation. Results from cross-national research conducted by Ergas and York (2012) demonstrate that CO_2 emissions per capita are lower in nations where women have higher political status. As an example, Iceland is considered the most gender-equal nation, with 48% women in parliament in 2023, and produced 3.9 metric tons of CO_2 emissions per capita in 2020. For comparison, the US's per capita emissions were at 13 metric tons per capita in 2020 (Statista 2023). Nations with greater women's representation in governing bodies have lower climate foot-prints. Put simply, gender equity in education and decision-making bodies, such as local councils and governments, leads to better climate outcomes.

These results demonstrate the need for more gender equity and democracy for a healthier planet. Promoting women's education and leadership is essential to miti-gating climate change.

Toward a culture of care

A central argument for ecofeminist scholars is that there is a core logic of domi-nation underlying the global economy and its supporting culture that legitimizes exploitation and violence. This logic of domination has roots in the philosophies of the Scientific Revolution that informed the colonial expansion of capitalism. To counter the harm that this codified logic perpetuates, activists and scholars argue for a cultural transition toward a historically feminized narrative of care. Priorities would shift from valuing profit above all else to valorizing care work, nurturing indi-vidual and collective well-being, and ecological restoration. We need a culture of radical sustainability, one that is at once socially and ecologically transformative – dismantling hierarchies toward total liberation – and regenerative – healing and restoring the health of people and the planet (Ergas 2021).

Research strongly suggests that caring for those at the bottom of our social hier-archies serves to lift everyone else up in turn, including degraded ecosystems (Pel-low 2014). Social transformation is a large-scale means of social change aimed at dismantling social hierarchies toward an equitable system.

Feminist activists and scholars have called the politics or ethic that should guide this transformation the "care ethic" or the "love ethic." bell hooks (2000:87) asserts that "the underlying values of a culture and its ethics shape and inform the way we speak and act." She proposes we move toward a love ethic, because "awakening to love can happen only as we let go of our obsession with power and domination" (p. 87). She further acknowledges that "embracing a love ethic means that we uti-lize the dimensions of love – 'care, commitment, trust, responsibility, respect, and knowledge' – in our everyday lives" (p. 94). Engaging in the ethic of care means that we attempt to leave things better than we found them. From the care ethic, we can begin to restore, revitalize, and regenerate local communities and ecologies, and eventually, the biosphere. But this ethic must first imbue the stories and larger cultural narrative with care. Rather than believing that humans own and control nature, we should turn our focus toward reciprocity and mutual care to heal.

An example of how to move toward reciprocity is by advancing community care. As activist Nakita Valerio explains,

[C]ommunity care is focused on the collective: taking care of people together, for everything from basic physical needs to psychological and even spiritual ones. Community care is a recognition of the undeniable cooperative and social nature of human beings and involves a commitment to reduce harm simply through being together . . . this might mean receiving messages from someone who needs to be comforted and heard, bringing dinner to a sick friend or packing up an abused friend's belongings as part of their exit plan from domestic violence.

(Valerio 2019)

Other aspects of community care may include spaces to process community-wide grief.

Systemic forms of oppression cause community-wide mental and physical harm and thus require collective approaches to healing (French et al. 2019). Radical healing is part of a process that centers on the strengths and resilience of oppressed communities. It first calls for developing an awareness of systemic forms of oppression; allowing this consciousness to inform action against institutionalized violence; and finally, proactively working to prevent recurring trauma. Collective grieving and healing are also necessary for people fighting for climate justice. Some organizations, such as Good Grief and 350.org, are taking on collective healing efforts, especially in relation to climate change, by providing support groups and ten-step programs that help people move beyond debilitating grief toward action (Good Grief Network 2019). While collective healing is crucial within communities affected by institutionalized violence, action is needed beyond internal community work, as external factors will continue to affect them. This external work can be conceived of as a form of regenerative community care, whereby allies and accomplices share in the responsibilities of recognizing systemic forms of oppression, educating others, and fighting against them.

Collective action is a means of fighting against systemic forms of oppression. While women are not as prevalent in mainstream, professional environmental movements, more women than men around the globe tend to lead and participate in grassroots environmental justice movements (Bell and Braun 2010). Women cite their roles as caretakers as their primary reason for starting or joining existing movements. They report engaging in this activism to protect their families and their communities from resource scarcity, natural disasters, and environmental toxins that threaten their health and livelihoods (Bell 2013). An example is the Chipko, which means to embrace, movement in India. In the 1970s, timber production in the Himalayan region created resource scarcity (e.g., fuelwood) as well as soil erosion that caused disasters, such as flooding and mudslides. In response, women from certain villages embraced trees in surrounding forests to prevent loggers from cutting them down. This movement was successful in changing regional laws to prevent logging in certain areas (Mies and Shiva 2014). Similar environmental justice movements exist in the Global North. Specifically, in the United States, working-class women in Appalachia largely spearhead the movement against mountaintop removal, despite many obstacles, including death threats (Bell 2013). Social movements are a necessary part of the process toward total liberation.

Conclusion

Climate change is a result of a culture that emphasizes exploitation and extraction. Western societies have championed ideas of domination since the Scientific Revolution. These ideas extend to the treatment of women as well as the planet. As such, an environmental revolution must also center the empowerment of and equity for women. Indeed, as discussed, inequality for women is associated with myriad forms of environmental harm.

A cultural shift that empowers women and centers on mutual care is necessary to prevent the worst results of the climate crisis. Women's education and associated involvement in political processes and organizing result in better environmental outcomes. Changing our cultural values from those centered on exploitation and wealth attainment to ones emphasizing mutual care and love is our only hope.

Note

1 Dominionism refers to Judeo-Christian beliefs in humans' God-ordained superiority over all other species and humans' dominion over the earth and its resources. It is believed that God created the earth for human use and prosperity.

References

Agyeman, Julian, Robert Bullard and B. Evans. 2003. "Joined-Up Thinking: Bringing Together Sustainability, Environmental Justice and Equity." In *Just Sustainabilities: Development in an Unequal World*, edited by Julian Agyeman, Robert Bullard and B. Evans, 1–16. London: Earthscan.

Bell, Shannon Elizabeth. 2013. *Our Roots Run Deep as Ironweed: Appalachian Women and the Fight for Environmental Justice*. Chicago and Urbana: University of Illinois Press.

Bell, Shannon Elizabeth, Braun, Yvonne. 2010. "Coal, Identity, and the Gendering of Environmental Justice Activism in Central Appalachia." *Gender and Society* 24(6): 794–813.

Brown, Alleen. 2018. "Five Spills, Six Months in Operation: Dakota Access Track Record Highlights Unavoidable Reality – Pipelines Leak." *The Intercept*. Accessed November 15, 2023. https://theintercept.com/2018/01/09/dakota-access-pipeline-leak-energy-transfer-partners/#:~:text=Since%20the%20leak%20was%20contained,least%20five%20times%20in%202017

Cannon, Terry. 2002. "Gender and Climate Hazards in Bangladesh." *Gender and Development*, 10(2): 45–50.

Denton, Fatma. 2002. "Climate Change Vulnerability, Impacts, and Adaptation: Why Does Gender Matter?" *Gender and Development*, 10(2): 10–20.

Dietz, Thomas, Eugene A. Rosa and Richard York. 2007. "Driving the Human Ecological Footprint." *Frontiers in Ecology and the Environment* 5(1): 13–18.

Ergas, Christina. 2014. "Barriers to Sustainability: Gendered Divisions of Labor in Cuban Urban Agriculture." In *From Sustainable to Resilient Cities: Global Concerns and Urban Efforts. Vol. 14*, edited by William G. Holt, 239–263. Bingley, UK: Emerald Group Publishing Limited.

Ergas, Christina. 2021. *Surviving Collapse: Building Community Toward Radical Sustainability*. New York: Oxford University Press.

Ergas, Christina, Patrick Greiner, Julius McGee and Matt Clement. 2021. "Does Gender Climate Influence Climate Change? The Multidimensionality of Gender Equality and Its Countervailing Effects on the Carbon Intensity of Well-Being." *Sustainability* 13(7): 3956. https://doi.org/10.3390/su13073956

Ergas, Christina, Laura McKinney and Shannon Bell. 2021. "Intersectionality and the Environment." In *International Handbook of Environmental Sociology*, edited by Beth Schaefer Caniglia, Andrew Jorgenson, Stephanie A. Malin, Lori Peek, David N. Pellow and Xiaorui Huang, 15–34. Switzerland: Springer.

Ergas, Christina and Richard York. 2012. "Women's Status and Carbon Dioxide Emissions: A Quantitative Cross-national Analysis." *Social Science Research* 41: 965–976.

French, Bryana, Jioni Lewis, Della V. Mosley, Hector Y. Adames, Nayeli Y. Chavez-Dueñas, Grace A. Chen and Helen A. Neville. 2019. "Toward a Psychological Framework of Radical Healing in Communities of Color." *The Counseling Psychologist* 1–33.

Gaard, Greta. 1997. "Toward a Queer Ecofeminism." *Hypatia* 12(1): 114–137.

Gaard, Greta. 2017. *Critical Ecofeminism*. New York: Lexington Books.

Good Grief Network. 2019. "10-Steps to Personal Resilience and Empowerment in a Chaotic Climate." Accessed November 15, 2023. www.goodgriefnetwork.org/

Green Belt Movement. 2023. "Our History. The Green Belt Movement." Accessed November 15, 2023. www.greenbeltmovement.org/who-we-are/our-history

Harvard Law. 2023. "The Dakota Access Pipeline (DAPL)." Accessed November 15, 2023. https://eelp.law.harvard.edu/2017/10/dakota-access-pipeline/

hooks, bell. 2000. *All About Love: New Visions*. New York: Harper Perennial.

Kimmerer, Robin Wall. 2013. *Braiding Sweetgrass: Indigenous Wisdom, Scientific Knowledge, and the Teachings of Plants*. Minneapolis: Milkweed.

Kwauk, Christina and Amanda Braga 2017. "Brooke Shearer Series: Three Platforms for Girls' Education in Climate Strategies." Global Economy and Development at Brookings. Accessed March 27, 2020. www.brookings.edu/research/3-platforms-for-girls-education-in-climate-strategies/

Lorber, Judith. 1994. "Night to His Day: The Social Construction of Gender." In *From Paradoxes of Gender*, 32–36. New Haven, CT: Yale University Press.

Lutz, Wolfgang and Samir Kc. 2011. "Global Human Capital: Integrating Education and Population." *Science* 333(29): 587–592.

Merchant, Carolyn. 1990. *The Death of Nature: Women, Ecology and the Scientific Revolution*. New York: Harper One.

Mies, Maria and Vandana Shiva. 2014. *Ecofeminism*. New York: Zed Books.

Monet, Jenni. 2016. "For Native 'Water Protectors,' Standing Rock Protest Has Become Fight for Religious Freedom, Human Rights." PBS. Accessed November 15, 2023. www.pbs.org/newshour/nation/military-force-criticized-dakota-access-pipeline-protests

Natural Resources Defense Council. 2018. "Flint Water Crisis: Everything You Need to Know." Accessed November 19, 2023. www.nrdc.org/stories/flint-water-crisis-everything-you-need-know#update

Pellow, David N. 2014. *Total Liberation: The Power and Promise of Animal Rights and the Radical Earth Movement*. Minneapolis: University of Minnesota Press.

Rocheleau, Dianne, Barbara Thomas-Slayter and Esther Wangari. 1996. *Feminist Political Ecology: Global Issues and Local Experiences*. New York: Routledge.

Schmuck, H. 2002. "Empowering Women in Bangladesh." International Federation of Red Cross and Red Crescent Societies. Accessed August 20, 2018. https://reliefweb.int/report/bangladesh/empowering-women-bangladesh

Shandra, John M., Carrie L. Shandra and Bruce London. 2008. "Women, Non-Governmental Organizations, and Deforestation: A Cross-National Study." *Population and Environment* 30(1–2): 48–72.

Statista. 2023. "Share of Women in the Icelandic Parliament from 1974 to 2021." Accessed November 17, 2023. www.statista.com/statistics/1090585/share-of-women-in-the-icelandic-parliament/

Thomas, William Isaac and Dorothy Swaine Thomas. 1928. *The Child in America: Behavior Problems and Programs*. New York: Alfred A. Knopf.

United Nations Development Program. 2007/2008. *Fighting Climate Change. Human Development Report*. Accessed July 12, 2010. hdr.undp.org/en/media/HDR_20072008_EN_Complete.pdf.

Valerio, Nakita. 2019. "This Viral Facebook Post Urges People to Rethink Self-Care" *Flare*. Accessed July 8, 2019. www.flare.com/identity/self-care-new-zealand-muslim- attack/

Wilson, K. 2017. "In the Name of Reproductive Rights: Race, Neoliberalism and the Embodied Violence of Population Policies." *New Form* 91: 50–68.

York, Richard and Christina Ergas. 2011. "Women's Status and World-System Position: An Exploratory Analysis." *Journal of World-Systems Research* 17(1): 147–164.

PART 2

Recognizing existing use of alternative stories

4

DOING ONE'S PART OF THE JOB

The Norwegian dugnad tradition in a global climate perspective

Anne Kristine Haugestad and Kari Marie Norgaard

The problem: Despite widespread knowledge of climate change, no nation or community has effectively curbed their emissions of climate gases. At the individual level, many people describe a sense of being gripped by fear of the future and overwhelmed by the scale of the problem. Cultural norms of what to feel or think leave little room for honest dialogue about uncertainty, while social constraints on action curb community-minded action. Capitalism sets institutional logics that incentivize choices and behaviors in the wrong direction.

The solution: Cultural tools can facilitate engagement. In this chapter, we describe how the Norwegian notion of "dugnad" (shared purpose for collective action) can be used to counter the sense of overwhelm individuals have described to create positive social pressures for collective engagement. We present a neighborhood scheme of cooperation that has functioned for more than 800 years as a mechanism to support collective action. The dugnad framework can be applied to communities and larger institutions alike. Adding a sociological understanding to how social change happens can upscale this effort even more.

Where this has worked: The Norwegian concept of dugnad (and similar contexts and attitudes in other Scandinavian countries) has long mobilized collective community actions. We propose that dugnad can be applied to mobilize climate engagement as well.

DOI: 10.4324/9781003437345-7

Despite widespread knowledge of the profound threat our changing climate poses to humanity and life on earth, the emissions of climate gases have continued to rise. To change our trajectory, numerous social processes and aspects of our economies and individual lifestyles must be decarbonized (Cho 2022).

While there is some evidence that transformations in patterns of consumption and production are emerging, these are not taking place quickly enough. At the individual level, many people describe a sense of being gripped by fear of the future and overwhelmed by the scale of the problem. As sociologists we understand that what an individual does, feels, or thinks about, is shaped by the "common sense" of their community, the constraints of normalcy, and the political and economic prospects and limitations. Cultural norms of what to feel or think leave little room for honest dialogue about uncertainty, while social constraints on action curb community-minded action. Capitalism sets institutional logics that incentivize choices and behaviors in the wrong direction.

Among the challenges that make responding to climate change difficult are the unprecedented challenges to the human imagination. If you want to solve a problem, it matters how you understand it. How do we imagine the reality of what is happening to the natural world? How do we imagine the relationship between any individual action, like driving a car, and climate impacts such as increased frequency of forest fires? One barrier to effective climate response has been the difficulty people have in visualizing our impacts on such a large system as the global climate. People need the ability to perceive the relationships between human actions and their effects on earth's biophysical systems – call it an "ecological imagination." Carbon footprint calculators have been very useful in assisting people to visualize that our various carbon-generating actions have "real" impacts on the atmosphere, even when people cannot directly see them. Yet, the quality of visualization that carbon footprint calculators provide is both misleading and insufficient, because they lead people to visualize their actions in individual terms, masking the societal factors that make huge variations in the emissions from a given hot shower or a cup of coffee.

But most importantly, how do we imagine which small-scale or individual actions can matter for changing course on such a global problem? The individualistic frameworks of carbon calculators have not been enough to leverage larger-scale social change. Considering the complexity of climate change as a global problem, we need ways to visualize how our actions can matter to heal the climate. We need to be able to see the relationships within society that make up this environmentally damaging social structure. This second form of visualization is about understanding how and under what circumstances social actions can matter on the global scale (Norgaard 2011; 2018).

In our research on Norwegians' attitudes and feelings related to climate change and environmental justice, many people described a desire to do one's part of the job to save the climate. One woman described how "It's terrible to think that we live so well while others live in such miserable circumstances." We also encountered frustration about what can be done and what level of responsibility should be taken for a global problem. This was the case for a high school student who noted, "I think there are a lot of people who feel, no matter what I do I can't do anything about that anyway." In addition to an ecological imagination – here linked to the understanding of what a carbon footprint expresses – there is a need to foster a "sociological imagination" on how the collective actions of many individuals can generate substantial change. Cultural and institutional mechanisms are needed to reshape the social landscape and leverage conditions for engagement. Our individualist Western societies have fewer and fewer mechanisms for collective action. How can people move beyond a sense of fear or guilt when facing climate change and become more engaged? How can social norms that currently inhibit action be reframed to support climate engagement? Even within modern capitalist societies, long-standing traditions that go beyond individualistic social relations and encourage people to work together for the collective good still persist. Work on "informal economic activity," ranging from farmers' markets, barter systems, or shared childcare, describes how individuals and communities sustain life outside of the capitalist market (Federici 2019), survive economically, and resist the logics of capitalist life (Gómez-Barris 2017; Linebaugh 2009). Can such "traditional" remnants of precapitalist social relations also serve as structures to mobilize climate engagement?

One way to understand culture is to think of it as a tool kit that contains a particular set of tools available as cultural resources. In her groundbreaking 1986 article, Anne Swidler argued that "culture influences action not by providing the ultimate values toward which action is oriented, but by shaping a repertoire or 'tool kit'" (1986:273). Swidler argued that culture shapes social action not by providing guiding values but by providing cultural components or "chunks of culture" (283) that can be used as tools by individuals to construct "strategies of action" (273). Each culture has a "tool kit" containing "symbols, stories, rituals and worldviews which people may use in varying configurations to solve different kinds of problems" (273). The purpose of our chapter is to give a description of a potential mechanism for collective action rooted in the traditional Norwegian practice of "dugnad" or shared purpose for collective action, and to suggest that this scheme of cooperation can be applied to shared efforts to save the climate.

Norgaard's (2011) work identifies the desire to avoid negative emotions such as fear of the future, guilt, and a sense of helplessness as part of what creates a sense of paralysis. Here psychological forces – including the need for a positive sense of individual and national identity, a sense of security, and a sense of self-efficacy – fly in the face of each of these emotions, reducing people's desire to engage. Using Swidler's tool kit metaphor, we can understand the practice of dugnad as a culturally meaningful "tool" people can use to respond to the problem of how to act on the climate. Here we describe how the notion of "dugnad" can be mobilized to counter both the sense of overwhelm many have described and to create positive social pressures for collective engagement. We present a neighborhood scheme of cooperation that has functioned for more than 800 years as a mechanism to support collective action.

According to the dugnad tradition, everybody in a neighborhood or an organization is supposed to contribute their time and work toward commonly defined tasks (Lorentzen and Dugstad 2011; Simon and Mobekk 2019). The dugnad framework can be applied to communities and larger institutions alike. Community members may come together to work on an issue in the local school, rebuild a common road, or write letters in support of political prisoners around the world. Organizations can use the notion of dugnad to justify actions for collective climate good to their members. Adding a sociological understanding to how social change happens can upscale this effort even more. While our focus is on generating individual and community mobilization, this scale of action must be integrated with institutional-level energy transitions (McGee and Greiner 2018), industry sector decarbonization (Rajabloo et al. 2022), and adapting to coming downshifts together with which any model of community mobilization must be combined.

Materials and methods

Our interview and observational data on how people in Norway experience and think about climate change and other global issues have been collected under two different contexts, and by social scientists with different theoretical and national backgrounds (one of us is Norwegian, the other American). We share a fascination with how people make sense of larger social and environmental issues of our time and a political desire of creating a world of social and environmental justice.

One data set (that of Kari Marie Norgaard) comes from an ethnographic analysis of public perception of climate change in a rural community known as "Bygdaby" during the winter of 2000–2001 (Norgaard 2011). This research involved 9 months of ethnographic observation, media analysis and semistructured interviews with 46 community members. Interview questions from this study were directed at how people perceived and made sense of the issue of global warming and other issues of global environmental justice, as experienced in the context of their local community.

Anne Kristine Haugestad describes the second data set as an exploration of potentials for overlapping consensus on global climate justice (Haugestad 2004). This material consists of 28 qualitative interviews on "consumption and distribution in today's world." These interviews were conducted in 2002 with Norwegian grassroots politicians from 15 political parties, which represent a broad political spectrum.

Twenty years has passed since we collected our data, but the Norwegian political discourse about the role of citizen-consumers (people who make politically conscious purchasing decisions) in preventing dangerous climate change is still almost nonexistent. Our results are still relevant in the Norwegian context, as well as for concerned citizen-consumers all over the world. When it comes to the descriptive dugnad ethic that becomes visible in our interview material, this is a deeply rooted cultural trait that is not subject to rapid changes (Simon and Mobekk 2019).

Despite different scopes and methodologies, similar stories about Norway, Norwegians, and the global challenges they face emerge from both studies. The people with whom we spoke described a series of unpleasant emotions that arose when they tried to think about acting on behalf of the climate. These included a sense of climate change as hard to really imagine, a sense that it was unpleasant to think about because it was overwhelming and made them feel guilty, and articulations of the collective action problem ("why should I bother when so many others do nothing?"). Tensions between local and national welfare concerns and global environmental concerns prevailed, but the people we interviewed also described continuing concern for global environmental problems and a sense that they wanted to engage, even as they may have been unsure of what to do. In the next section, we share examples of each experience in respondents' own words, and then introduce the Norwegian notion of dugnad, showing how it can be a powerful tool for encouraging people to act in the face of these particular constraints.

Care for one's own as well as the global neighborhood?

In her interviews with people in Bygdaby, Norgaard found that while information was *known* on some level, it was not necessarily being integrated into immediate reality. Liv, a woman in her late 20s who was involved in human rights work in her community, explained how "you had the knowledge but lived in a completely different world":

> You have the knowledge, but you live in an entirely different world. That's why I say you just flat out have to sort. And not take everything in. It's not possible to go around thinking about it the whole time. You would give up, right? We sit here so safe and good, even if there are bad things that are happening here too – it's like nursery school if you start comparing it with some places. Just think of our neighbor, Russia, how things are there!

In Liv's words we see evidence that part of what is hard about becoming engaged is that it is hard to even imagine present reality. Instead, Liv and others actively disconnect what they know and their daily lives. This sense of "living in an entirely different world" because "it is not possible to go around thinking about it all the time" was echoed by others such as Kjersti, a teacher at the high school:

> We live in one way, and we think in another. We learn to think in parallel. It's a skill, an art of living.

Kjersti describes the ability to "think in parallel" as a skill or art that both comes from and is necessitated by everyday life. Since members of the community did know about global warming but did not integrate this knowledge into everyday life, they experienced what Lifton calls a state of *double realities* (1982). In one reality was the collectively constructed sense of normal everyday life. In the other reality existed the troubling knowledge of increasing automobile use, polar ice caps melting, and the predictions for future weather scenarios.

A second theme people with whom we spoke described was a norm conflict between care for one's own household and global care. If individuals are invited to take part in a climate dugnad, this might imply sacrifices at the household level. They might feel that they are expected to give up too much comfort and life quality. If other individuals are not committed, such personal sacrifices will have insignificant global impact. In our material, resistance against playing a role as "forerunner consumers" in a global effort to save the climate often comes to the surface in references to the American level of consumption. Why should Norwegians sacrifice welfare when Americans don't?

While the purchase of "green" consumption goods can be done without much harm to family welfare, we have quite a different situation when consumption of energy and fossil fuels is included in the picture. If the efforts to curb carbon dioxide (CO_2) emissions are pictured as a job to which decent people should contribute voluntarily, Norwegians are certainly asked to make sacrifices for the global common good. For example, Eirik, an environmentally active local political leader from Bygdaby, the community of focus in Norgaard's work, describes the difficulty of living by his conscience:

> We shouldn't consume so many resources. Drive so much, or travel so much by air. We know that it is bad because it increases CO_2 levels. And creates a worse situation. But at the same time of course we want to go on vacation, we want to go to the South, we want to, well, live a normal life for today. So many times I have a guilty conscience because I know that I should do something, or do it less. But at the same time there is the social pressure. And I want for my children and for my wife to be able to experience the same positive things that are normal in their community of friends and in this society. It is very . . . I think it is a bit problematic. I feel that I could do more, but it would be at the expense of, it would perhaps create a more difficult relationship between me and my children or my partner and in general. It really isn't easy.

Eirik expresses guilt in the context of social relationships. He describes how it is in part his connection to others in his community that makes taking action difficult. Although privileged people around the world experience this contradiction between their wealth and the poverty of others, there is a particular force in the way in which these issues came together for people in Bygdaby. High levels of access to information, high levels of acceptance of the information, a strong tradition of values for social welfare and environmental consciousness, and current wealth and current economic interests came together with force (Norgaard 2011).

Given their high newspaper readership and level of knowledge about the rest of the world, community members were aware that their economic situation and standard of living was better than what people in most other countries can hope for. This understanding contrasted sharply with their desire for and values of equality and egalitarianism, thus raising feelings of guilt.

Several interviewees in Bygdaby explain that they can't care too much about global problems, because that will make them feel miserable. They say that they "have to protect themselves a little." Heidi, one of the grassroots politicians in Haugestad's study, describes a "door" that must be kept closed:

> I do have a guilty conscience because my consumption is so high, because I don't drive an electric car. Of course, I use too much of everything. This has more to do with the environment. If one also introduces issues of poverty when considering one's consumption, . . . then one would be *breaking into pieces*. . . . When you see those slaves . . . you are just . . . At some level you know, when you put on your Nike shoes . . . then you know, if you open that door, then you know that someone has lived as slaves to produce them. So, it's a guilty conscience that one somehow tries to control by not thinking about it, and one says to oneself: "Why shall you take everything so damn seriously?" . . . And it's very uncomfortable; it's a very uncomfortable way of life. Because you are so woven into such a . . . such a *viciously* unjust system.

Heidi's description of her situation supports an interpretation of the Norwegian situation as being characterized by lack of manageable tasks, not lack of care. Many Norwegians are mobilized to want to participate in the efforts to achieve a more resource-friendly and sustainable consumption pattern. But at the same time, one doesn't want to sacrifice one's own or the family's welfare if it's not actually part of a joint effort where one's own sacrifices make sense. This situation might be interpreted as collective powerlessness: Many people want to contribute, but it seems meaningless to sacrifice one's own welfare if no one else does it. There seems to be a lack of a sense of a common narrative and plan of action that might transform all the mobilized willingness to consume more responsibly into actual changes in resource use.

The dugnad: a powerful tool for collective action

The word "dugnad" tells people that they are expected to do their part of a defined joint effort: "Say dugnad, and without further explanation people will gather for a joint big effort," Lorentzen and Dugstad write in their book about the Norwegian dugnad tradition (2011:9).[1] The essence of the dugnad is "*a we feeling* among a group of people which motivates to collective work effort for a goal that concerns all the participants" (Lorentzen and Dugstad 2011:12). The noun "dugnad" comes from the verb "duge," which means to be fit or to be good enough (Simon and

Mobekk 2019:817). The adjective "dugelig" is often used to describe a person who does a good job.

The dugnad tradition dates more than 800 years back in time and probably started as obligatory schemes of cooperation to solve tasks where many hands were needed (i.e., the landing of big boats) (Lorentzen and Dugstad 2011:18–19). From the beginning of the 17th century, the word "dugnad" became the established term for different forms of exchange work (Lorentzen and Dugstad 2011:21). As exchange work the dugnad was never statutory, but we know of its existence from mentions in documents from court proceedings because there have been fights and turmoil during social gatherings that traditionally follow when a dugnad job has been finished. So important is the concept of dugnad to Norwegian identity that in 2004 it was declared "the Norwegian national word."

For centuries, dugnads were primarily about many people helping one man or one farm or family with a job where many hands were needed to finish the job within a limited time frame. Putting a new roof on a farm building was a typical dugnad job (Norddølum 1980:105). In a rural neighborhood, the dugnad system could even fulfill some of the functions of modern insurance and the welfare state. If a farmer experienced a fire or was behind with normal farm work because of illness, neighbors could gather for a "helping dugnad" (Klepp 2001:83). In parts of Norway, the original version of the dugnad lived on well into the 20th century, but inclusion in the market economy meant that farmers increasingly bought helping hands instead of gathering neighbors for a dugnad (Klepp 1982).

While the significance of the original form of a dugnad – many people helping one man or one farm or family, followed by a meal and perhaps even a big feast – decreased, another form of dugnad gained significance with the emergence of voluntary organizations from the 1840s and onward. This form of the dugnad is about many people creating a common good together. The common good could, for instance, be an assembly hall for a voluntary organization or a playground for a neighborhood or housing association. Lorentzen and Dugstad (2011:13) list the following core elements of such community dugnads:

- Unpaid work: No one receives payment for what they do.
- Simultaneity: People meet face-to-face.
- Work: Those who come together perform joint work tasks.
- Time limit: A defined beginning and end (usually one or two days).
- Social elements: Party, meal, or other activities that increase the participants' community feeling.

For neighborhood groups, the dugnad creates a space where members of different households can meet and create a nice and safe neighborhood even if they do not have much more than the neighborhood in common. The dugnad leaders are in charge of pointing out suitable dugnad tasks and supervising the work if necessary. The dugnad opens for different kinds of constructive activities, and a good dugnad leader allows the dugnad participants to work as independently as possible and to choose between several tasks.

In recent decades, the word "dugnad" has come to be decoupled from cooperation within a specific neighborhood or organization and can now be used about any *big job* that must be done within a quite *short time frame*, and with results that *benefit the society as a whole*. The authorities can try to invoke "the dugnad spirit" when they want people "to tolerate something that requires a certain degree of self-sacrifice" (Klepp 2001:90).

The Norwegian dugnad has thus survived and developed as a grounded tool for collective action which can now be used in any constructive effort that benefits a community or the society as a whole – such as a Global Climate Dugnad. As social scientists, we understand that we need mechanisms or tools that change the social reality (i.e., how people perceive potentials and hindrances in the social world), and we believe that the concept of "dugnad" can help people understand that their actions matter.

Our data sets make a grounded *descriptive* dugnad ethic visible: Norwegians tend to expect themselves and others to do their part of a collective effort that needs to be done, and they feel bad if they do not see how they can in fact do their part of the job. When applied to the global climate challenge, the dugnad ethic does however become a *prescriptive* ethic: All world citizens must do their part of the job to save the climate.

The Global Climate Dugnad as convergence to equal shares per person globally

In this chapter, we define the job that must be done as a steady decrease in the carbon footprint of all world citizens who have a carbon footprint that is bigger than a sustainable global average. There are infinite ways that these outcomes can be achieved. By average, a world citizen has a carbon footprint amounting to 4.5 tons CO_2 per year, and a sustainable global average is below 2 tons CO_2 per year (World Bank 2023; The Nature Conservancy 2023). The Norwegian average is approximately 11 tons CO_2 per year, while the number for the richest 10% in the European Union is 20 tons CO_2 per year. The richest 1% in Luxembourg have a carbon footprint amounting to 214 tons CO_2 per year, while the number for an average Ethiopian is 0.1 tons CO_2 per year (Riise 2021:19–25). A decrease to a sustainable carbon footprint thus implies a massive decarbonization of Western lifestyles and societies (Cho 2022).

In Haugestad's interviews with grassroots politicians, the interviewees were introduced to carbon footprint calculations as a tool to measure an individual's contribution to the global efforts to save the climate – with equal carbon footprints at around 1 ton per world citizen as the long-term goal. Many of the grassroots politicians immediately embraced this tool as a potential framework for globally responsible local living. The reason for this enthusiastic response is probably that the tool models responsible consumption as a clearly defined task.

If information on the carbon footprints of goods and services is universally available, and each person automatically gets a record of their personal carbon footprint, this creates an arena – or a playground – for the Global Climate Dugnad to take place. The task is simple – to steadily decrease one's carbon footprint. Ulf, who is one of the grassroots politicians in Haugestad's study, initially was quite skeptical about the idea to register each person's carbon footprint, but later he became more positive because he could see that this could be turned into a game or a sport.

Another grassroots politician, Gerda, was very interested in learning about how she could act climate friendly in her everyday life. Through the interview, Gerda became an extreme supporter of carbon footprint calculations, and she fully endorsed a steady decrease in one's carbon footprint as a way to free oneself from guilt about one's own privileges compared to most of her global neighbors. She wanted an easy way to keep track of her carbon footprint as soon as possible.

Simon and Mobekk (2019) write about how the existence of the dugnad contributes to Norwegians being socialized into prosocial attitudes and behavior. One of Haugestad's findings is that the dugnad spirit is often present in peoples' reasoning without being mentioned. None of the interviewees in Haugestad's study explicitly referred to the dugnad scheme of cooperation, but all of them answered in line with this implicit ethic when they talked about the climate challenge. When analyzing the interview material, Haugestad looked for the political cause closest to each interviewee's heart. An older woman, Agnes, had become involved in politics because she was very angry at politicians. She meant that politicians were destroying Norway. At first Haugestad couldn't identify any political cause that was close to Agnes's heart, but then a statement stood out:

> It would be best if everyone is proper in their behavior and does not act as parasites and swindlers. If everyone could be proper, things would be better.

This statement made Haugestad realize that Agnes was a fierce protector of the dugnad culture where everyone does their part of the job rather than being a free rider and benefiting from the system while others suffer.

Conclusion: the power of a Global Climate Dugnad

The tradition of dugnad combined with a clear picture of the job that must be done to save the climate seem to have a potential to convert frustration into action and enterprise. When responding to the climate challenge is modeled as manageable dugnad tasks, this can mobilize people's wish to do their parts of the job. If the tasks are manageable, there seems to be a potential for such mobilization even if there is no guarantee that other people will do their part of the job. Dugnad is a tool that fixes the problem of the insufficient scale of individual action, the paralysis people face in wondering why they should act when others do not. Working together through a Global Climate Dugnad provides a culturally relevant ethical imperative for action and scales up the power of individual actions. Doing one's part of a defined job seems to be intrinsically meaningful and serves as an antidote to feelings of guilt and helplessness. Furthermore, the dugnad ethic ties into long-standing cultural and social norms that encourage action at the same time as it promotes a positive sense of individual and national identity.

A dugnad is about creating a common good that benefits all the participants, so in principle no one should be left worse off after the dugnad than they were before it started. In today's world, it is obvious that if "worse off" is interpreted in purely economic terms, a global dugnad to save the climate is impossible. Some people in the world will have to accept some limits to economic privilege so that other people can get a decent share. If, however, we apply the multidimensional definition of welfare implicit in the sustainability agenda, a successful Global Climate Dugnad will in our opinion still make everybody better off. This multidimensional agenda includes peace, security, environmental sustainability, social sustainability, and life quality in addition to economic sustainability (World Commission on Environment and Development 1987). With such prospects, today's privileged world citizens might accept some limits to economic privilege.

When it comes to community dugnads, those who do not participate in the dug-nad are free to enjoy the results, but they must not destroy or exploit the goods created by the dugnad group. Applied to a Global Climate Dugnad, this implies that if one accepts the prescriptive dugnad ethic (to do one's part of a dugnad that has to be done), it is not really possible for an individual "overconsumer" to choose not to participate. The good created by the dugnad group is a reduction in climate gas emissions – and each of us must participate in this dugnad to save the climate. For the wealthy high-consuming citizens of places like Norway, there is no such thing as a neutral space outside the Global Climate Dugnad. It becomes an ethical imperative.

The chances for a change toward a global dugnad ethic to take place might seem microscopic. Strong actors benefit from the competitive world economy. But the same actors depend on the goodwill and trust of citizens, voters, and consumers. Citizen-consumers need help from businesses and politicians to be able to reduce their carbon footprint, and they might choose to support those businesses and politicians who provide such help. They can also request information on businesses' social and environmental performance. If they don't like what they find out – or if information is not made available – citizen-consumers can choose other businesses and politicians, who both in words and actions comply with the Global Climate Dugnad.

Individuals cannot solve climate change on their own, and they know it. Emissions reductions at the scale we need must be tackled at more collective levels. But the power of even smaller groups of people to create real change is another story. Small community groups can take on many types of meaningful actions. While we have focused on the Global Climate Dugnad to curb emissions globally, there are also many potential local dugnad tasks related to climate change. Such community climate dugnads are important both in their own right and as arenas for people to become aware of the potential for a Global Climate Dugnad. The list of potential community dugnads is infinite, and we can only mention a few of them:

- Learn about and provide input on community climate action plans.
- Write letters to politicians to create local laws to keep grocery stores from throwing away eatable food.
- Oppose new infrastructure for fossil fuel development.
- At grocery stores, ask about the carbon footprint of the food you consider buying. This can urge the shopkeepers to request such information from their suppliers.
- Carpool or join a car share instead of owning a car.
- Invite friends and neighbors to a dugnad to repair bicycles for people in your neighborhood.
- Use social media and other channels to create pressure for safe bicycle paths in your neighborhood.
- Invite friends and neighbors to a cooking course where you learn to make tasty dinners without meat.
- See if your local school, workplace, or church has a climate action plan. If so, read it and see how you can become involved. If not, work with your team to create one or help others to do so.
- Focus on a specific task within your local organization such as an energy audit, a transportation challenge, or setting up teams for a letter writing campaign.

Working together can scale up individual actions both through enhanced capacity for specific tasks and by the creation of a supportive community that can in turn provide hopeful action, generate political strategy, and provide capacity for facing fears about the future. Local, national, and global leaders should act as dugnad leaders and facilitators who make it as easy as possible for individuals and community groups to become responsible global neighbors by participating in both community climate dugnads and the Global Climate Dugnad.

We believe that the encouraging tendencies in our interview material can be linked to the dugnad scheme of cooperation that our informants seem to tacitly apply even to the global neighborhood. While we have provided an example specific to Norway, all nations and communities retain remnant social relations that may be meaningfully mobilized in this moment of climate crisis. While dugnad is a uniquely Norwegian cultural construct, Norwegian dugnad values – to do one's part of the job and expect others to do their part of the job, that is, to act according to the same principles that one would like others to act in accordance with – are not uniquely Norwegian. They are the values of all world religions (the golden rule), of the UN Declaration on Universal Human Rights, of Kant's categorical imperative, and of cooperating neighbors all over the globe. Many other cultures and traditions have practices about collective responsibility that could be usefully applied to a climate context as we have done here with the notion of dugnad. The African Ubuntu philosophy (propagating the distribution of wealth), the Finnish notion of "talkoot" (nearly mandatory, unpaid work for a common good), and many other cultural traditions of service, especially those involving community elements, can serve as cultural tools that simultaneously scale up climate actions, provide a reason to take action even in the face of uncertain outcomes, and provide the ethical norms to do so. As they do, they can be a means to generate positive local outcomes for individuals and communities while simultaneously producing climate benefits to the global scale.

Note

1 All quotes from Norwegian are translated by the authors.

References

Cho, Rene. 2022. "What Is Decarbonization, and How Do We Make It Happen?" *State of the Planet. News from the Columbia Climate School.* Retrieved November 22, 2023. https://news.climate.columbia.edu/2022/04/22/what-is-decarbonization-and-how-do-we-make-it-happen/

Federici, Silvia. 2019. "Social Reproduction Theory." *Radical Philosophy* 2(4): 55–57.

Haugestad, Anne K. 2004. "Norwegians as Global Neighbours and Global Citizens." In *Future as Fairness: Ecological Justice and Global Citizenship*, edited by A. K. Haugestad and J. D. Wulfhorst. Amsterdam and New York: Rodopi.

Klepp, Asbjørn. 1982. "Reciprocity and Integration into a Market Economy: An Attempt at Explaining Varying Formalization of the 'Dugnad' in Pre-Industrial Society." *Ethnologia Scandinavica. A Journal for Nordic Ethnology* 1982: 85–93.

Klepp, Asbjørn. 2001. "From Neighbourly Duty to National Rhetoric. An Analysis of the Shifting Meanings of Norwegian *Dugnad*." *Ethnologia Scandinavica. A Journal for Nordic Ethnology* 2001: 82–98.

Lifton, Robert J. 1982. *Indefensible Weapons: The Political and Psychological Case against Nuclearism.* New York: Basic Books.

Linebaugh, Peter. 2009. *The Magna Carta Manifesto: Liberties and Commons for All.* Berkeley: University of California Press.

Lorentzen, Håkon and Line Dugstad. 2011. *Den norske dugnaden. Historie, kultur og fellesskap* [The Norwegian Dugnad: History, Culture and Community]. Kristiansand: Norwegian Academic Press.

McGee, Julius Alexander and Patrick Trent Greiner. 2018. "Can Reducing Income Inequality Decouple Economic Growth from CO_2 Emissions?" *Socius* 4. Retrieved November 22, 2023. https//doi.org/10.1177/2378023118772716

The Nature Conservancy. 2023. "What is a Carbon Footprint?" Retrieved November 22, 2023. www.nature.org/en-us/get-involved/how-to-help/carbon-footprint-calculator/

Norddølum, Helge. 1980. "The 'Dugnad' in the Pre-Industrial Peasant Community. An Attempt at an Explanation." *Ethnologia Scandinavica. A Journal for Nordic Ethnology* 1980: 102–112.

Norgaard, Kari Marie. 2011. *Living in Denial. Climate Change, Emotions, and Everyday Life*. Cambridge, MA: MIT Press.

Norgaard, Kari Marie. 2018. "The Sociological Imagination in a Time of Climate Change." *Global and Planetary Change* 163. Retrieved November 22, 2023. http://dx.doi.org/10.1016/j.gloplacha.2017.09.018

Rajabloo, Talieh, Ward De Ceuninck, Luc Van Wortswinkel, Mashallah Rezakazemi and Tejraj Aminabhavi. 2022. Environmental Management of Industrial Decarbonization with Focus on Chemical Sectors: A Review. *Journal of Environmental Management* 302. https://doi.org/10.1016/j.jenvman.2021.114055

Riise, Anja Bakken. 2021. *Mitt klimaregnskap. Et forsøk på å leve bærekraftig og løsningene vi trenger* [My Climate Bookkeeping: An Attempt to Live Sustainably, and the Solutions We Need]. Oslo: Res Publica.

Simon, Carsta and Hilde Mobekk. 2019. "*Dugnad*: A Fact and a Narrative of Norwegian Prosocial Behavior." *Perspectives on Behavior Science* 42(4): 815–834.

Swidler, Ann. 1986. "Culture in Action: Symbols and Strategies." *American Sociological Review* 51: 273–86.

World Commission on Environment and Development. 1987. *Our Common Future*. Oxford: Oxford University Press.

World Bank. 2023. "CO_2 Emissions (Metric Tons Per Capita)." Retrieved November 22, 2023. https://data.worldbank.org/indicator/EN.ATM.CO2E.PC

5

IN/ACTION IN ADDRESSING THE CLIMATE CRISIS

The possibilities of Generation Z

Hannah Block, Cailin Lorek, and Ryan Alaniz

The problem: This chapter addresses the paralyzing impact of the dire messages received by Generation Z about the climate crisis.

The solution: Despite these messages, a subsection of Gen Z is committed to combating the climate crisis at the individual, cultural, political, and international arenas. To increase their political power and leverage these heroic efforts, we need to shift the cultural stories regarding the hopelessness of climate change to that of possibility. We need to "call in" inactive peers and older generations to join in the successful activism of Gen Z.

Where this has worked: In 2016, Ameriyanna "Mari" Copeny wrote a letter to former president Barack Obama about the Flint, Michigan, water crisis. Her letter motivated Obama to visit Flint and observe the situation for himself. Following the visit, he approved $100 million in relief for Flint (Copeny 2023). Greta Thunberg, a Swedish climate activist, organized a school strike for the climate, galvanizing millions of young people across the world.

The decade beginning in 2000 was the warmest on record (Earth Institute 2010), and natural disasters have nearly doubled in the last 20 years, largely in the form of climate-related disasters like floods, droughts, and storms (Larson 2020). In the past 20 years, we have experienced record-breaking heat waves, droughts, hurricanes, wildfires, and floods.

Gen Z (born between 1997 and 2012) compose more than 20% of the US population, but due to their young age, this generation has not had significant political power to tackle the challenge of climate change. This cohort was born into a

DOI: 10.4324/9781003437345-8

troubled world. Some young people feel the climate crisis is too big for any individual's actions to make a difference, while others are deeply engaged in implementing change. Indeed, Gen Z hosts the largest percentage of its members engaged in climate activism. For example, in 2021, approximately one-third of Gen Z in the United States noted that they have "done something in the past year to address climate change, such as donating money, volunteering, contacting an elected official, or attending a rally or protest" (Funk 2021). Unfortunately, the number of engaged young Americans is still insufficient to curb the negative human impact on the environment.

In a poll from *The Economist* and YouGov (2022), Americans were surveyed about various social issues, including their support of clean energy laws and the importance of addressing climate change. When asked about climate change, 80% of 18- to 29-year-old participants considered the issue important, which is higher than all other age groups. This demographic group also expressed that climate change was the most important social issue. They have contributed significantly less pollution than previous generations and yet they have been socialized to shoulder the responsibility of affecting change before the damage is irreparable.

Cailin's story

A member of Gen Z, I am perpetually torn between a determined belief humanity can act collectively to keep temperature levels below the two-degree rise threshold marked as significant by the United Nations and pessimistic despair about the challenges we will face in the next 50 years. I grew up in a household that valued engaging with the outdoors and in which climate change and global warming were often discussed.

In 2006, I sat in my elementary school auditorium and watched several young adults in brightly colored clothing dance and sing the tune "reduce, reuse, recycle!" This catchy phrase was meant to teach us to reduce our overall consumption, reuse items we already had, and recycle items such as glass and plastic. To this day that catchy song still lingers in my head. Other presentations consisted of tips to help mitigate the drought in California, during which we were told to turn off the water while we brushed our teeth and use our newly gifted low-flow shower heads. While this information was important to learn, it had the primary effect of scaring me.

In high school I attended an advanced placement environmental science class where I learned more about our contribution to carbon dioxide emissions. This knowledge led me to become a vegetarian for several years to mitigate my negative environmental impact. As a recent college graduate, I still eat a mostly plant-based diet, and I thrift most of my clothing rather than shopping online, but I am also aware that my individual choices are not enough to slow the warming of the planet, and pale in comparison to the negative consequences of corporations, institutions (such as the military), and nation-state in/actions. My feelings of helplessness reflect those of many other Gen Zers who struggle with the same dilemma.

This type of environmental education promoted individual behavioral changes by children as a solution to the global climate problem. By reinforcing the responsibility of the individual, the school diverted attention away from major polluters such as the oil and gas industry. Students were left feeling burdened to solve the climate crisis we did not create and yet told our only reasonable actions could be implementing simplistic strategies such as recycling (only about 5% of plastics are recycled) (Gammon 2022), carpooling, or taking shorter showers, none of which would significantly curb global carbon emissions.

Hannah's story

I have been passionate about the environment from an early age. I eagerly attended Earth Day celebrations, watched nature documentaries, and would excitedly fundraise by filling my "Save The Rainforest" cardboard box full of pennies for fundraisers. I loved the outdoors. I spent much of my playtime wandering through parks and admiring the flowers and trees, taking in their unique shapes and movements. I always wanted to be outside in some capacity. I had such an immense appreciation for nature and wildlife and had yet to learn about the fragility of the environment as I knew it.

As my educational horizons expanded, I began to better grasp the severity of the climate crisis and further explore the dire scientific and political perspectives on this issue. I started attending protests demanding for climate action and arguing with strangers online who denied global warming only to end up making no impact on the opinions of others and becoming angry at the overall lack of action both from the general public and from those in positions of power. During the 2020 COVID-19 lockdown, I spent more time online and became enveloped in negative news regarding climate change. This led to a sense of hopelessness about the state of the world and powerlessness as an individual to make a difference. The feeling was exacerbated by the presidential election. Hearing former president Donald Trump brush off data presented by climate scientists by saying "It will start getting cooler, you just watch" (BBC 2020) genuinely scared me. I cannot recall a time in my life in which I felt more demoralized than hearing the leader of our nation and the "free world" deny the existential threat of a warming planet.

Now, as a college graduate, I cannot help but picture what the rest of my life will look like. When I look at the future through a negative lens, I cannot help but wonder how many species will go extinct in my lifetime due to the collapsing environment or if I will be able to withstand the predicted rising temperatures when I am older. At the same time, I also try to look through a positive lens. I wonder if gasoline-powered cars and single-use plastic will stop being produced by my generation. I wonder if I will live to see the complete restoration of the ozone layer. I also wonder, if and when I have children of my own, if they will be able to have a childhood like mine; a childhood where they can experience nature and clean air and feel like they are part of their ecosystem, rather than the thing that is destroying it. These feelings coexist despite being contradictory. Although other generations may have a sense of this, the depth of hope and dread we feel is unique to Gen Z.

This chapter briefly reviews how Gen Zers have a paradoxical relationship with climate change – both fearing it and remaining apathetic about it – and how some have overcome apocalyptic prophecies to work for change. To do this we first review our culture's conflicting messages about global warming. We then describe the positive contributions that members of Gen Z are making to address environmental change on the micro- and macro-levels. Finally, we close with additional strategies to mobilize members of Gen Z to leverage their cultural and political power.

The climate messages taught to Generation Z

Growing up in the 2000s, members of Gen Z learned about the dangers of climate change in school but also by the media. They were taught about the impending doom of existential catastrophe and their responsibility as individuals to stop the crisis through their actions, such as turning off the lights when they leave the room. During their formative years, the fossil fuel industry giant British Petroleum (BP) developed the "carbon footprint," a public relations effort that placed the blame and responsibility for climate change on the individual rather than the broader structural issues that create the problem (Kaufman 2023). Even environmentally conscious organizations perpetuate the belief that individual actions could solve climate change. For example, the Nature Conservancy offers a short quiz that calculates an individual's carbon footprint and recommends lowering it through actions such as reducing meat consumption and line-drying laundry but fails to confront the structural issues of the oil and gas industry (The Nature Conservancy 2023).

In recent years, major news sources such as the BBC, CNN, CNBC, and *The New York Times* have all released articles advising readers on the best ways to reduce their carbon footprints. NBC, likely ignorant to the irony, took a consumerist approach with their article titled "Climate Neutral: How to shop for the best eco-friendly products" (Horvath and Pardilla 2021). Comparatively, these media outlets offer relatively little criticism concerning the major industries that most contribute to carbon dioxide emissions and fail to explore the potential for structural changes such as the development of mass transit, creating a meaningful carbon tax credit, or strengthening emission reduction commitments.

The erosion of hope

Being exposed to consistent reminders of the environmental crisis has caused many in this generation to express feelings of hopelessness, apathy, and fear. In 2009, *The Guardian* published an article titled "The environment in the decade of climate change" that describes how the world has "only a few years to avoid apocalypse" (Vidal 2021). Despite having prevented global catastrophe so far, the world continues to worsen in environmental crisis each year, leaving many young people feeling a sense of existential crisis (Ray 2020). Sociologist Kari Marie Norgaard – see Chapter 4 – describes how it is easier for many people to visualize and fear the end of the world than it is for them to envision a world where economic, cultural, and political structures shift to successfully combat climate change (Ray 2020).

This is the reality: after decades of activism, the world continues to warm at an unsustainable rate. This news creates a duality of distress and motivation. The inaction of previous generations and the obvious worsening of the planet serves to further anxiety and apathy for some while acting as a springboard for action by others.

It is also difficult to stay motivated when there are few clear environmental victories, resulting in a high level of burnout among justice-oriented youth. Many feel as though they need to sacrifice their own well-being to demonstrate their commitment to a movement. In her book *A Field Guide to Climate Anxiety: How to Keep Your Cool on a Warming Planet*, Sarah Jaquette Ray (2020) describes the influence climate change has had on some of her Gen Z students. One such student, Madi, describes how her environmental values led her to feel extreme guilt and self-loathing, ultimately harming her health. She "thought that 'to disappear, to become smaller, was to be beautiful' – and, of course, a good environmentalist" (Ray 2020).

Generation Z in action

Political change

Despite this discouragement, Gen Z contributes to climate change awareness and action at the individual, local, and national/international levels beyond that of other generations (Funk 2021).

Gen Z, though in some cases still too young to participate fully in the political process, is committing to political engagement at a higher rate than previous cohorts, including higher levels of civic engagement, demonstrating a shift in youth political involvement. For example, in 2020, 53% of young voters participated in the presidential election, as compared to only 36% in 2000 (Alfaro 2022; Jamieson 2002). In 2022, voter turnout for young Americans was the highest it has been in 30 years (Lopez 2022).

Gen Zers are organizing and building issue-driven political power – one of the most prominent of those is climate change. The Pew Research Center's survey, "How Americans' Attitudes About Climate Change Differ by Generation, Party and Other Factors," found that 67% of Gen Z adults believe the climate should be a top priority to protect the planet for future generations. Additionally, 37% said climate change was their top personal concern, and 32% said that they had taken personal action to combat the climate crisis within the last year. These numbers were higher than any other generation surveyed (Funk 2021). This may reflect young citizen activism to embrace political power through organizations such as Voters for Tomorrow and NextGen America. Both organizations utilize social media to reach a larger audience and encourage young people to register to vote. NextGen America alone has registered over 77,000 voters. Via TikTok videos, podcasts, digital outreach, in-person meet-ups, and organizations, young adults are finding their political voice.

Youth and young adults are increasingly involved in local and even national government, which not only has the power to make community-level changes but also shape important climate change mitigation and adaptation programs (Selin and VanDeveer 2009). For example, in Idaho, a deeply conservative state, an 18-year-old climate activist, Shiva Rajbhandari, was recently elected to the Boise school board (Mackey 2022). Farther east, Maxwell Alejandro Frost, a 25-year-old from Florida, became the first member of Gen Z elected to Congress. Representing Florida's 10th Congressional District, he attributes the climate crisis as one of the reasons he ran for political office. He explains, "We didn't cause the hurricane [sic] but science tells us that we are contributing to these more devastating effects. And so, the cost of not doing anything is far greater than the cost of taking bold action and this is one of the reasons why I decided to run for Congress" (Bruggers, Davis, and Dryfoos 2022). Like other Gen Zers, seeing the direct impact of the climate crisis on society simultaneously scared and inspired Frost to make a change.

Moreover, actions such as writing letters and attending town hall meetings to support specific people or propositions that combat global warming are productive ways in which Gen Z are using their voice and contributing to local change (Carnegie 2022). Perhaps most importantly, the United States continues to be a dominant player in international affairs; changes pushed up from the local to the national agenda will undoubtedly influence other nations and their policies and behaviors. Indeed, this can be seen in the historical movement to increase recycling. Earth Day, which originally began in the United States in 1970, called attention to the importance of recycling and is now celebrated in over 193 countries (Earthday.org 2024).

Beyond electoral politics, young people are bringing attention to the issue of a warming planet by attending protests and sharing content on social media (Funk 2021). Some are suing state governments to declare that their support of the fossil fuel industry is unconstitutional, as it robs young people of a healthy future. In *Held v. Montana*, the court sided with 16 youth activists, ranging in age from 2 to 18 years, that the state had an obligation to protect young people from Montana's greenhouse gas emissions and must find ways to curb emissions (Baker 2023).

Globally, youth are engaging in legal battles and winning in Colombia, the Netherlands, and Pakistan, among others (Parker 2019). Gen Zers have used nonviolent civil disobedience to bring public attention to the climate crisis and to compel governments to take action. These tactics include examples such as blockading the Federal Reserve to demand divestment in oil and gas companies, staging die-ins to draw attention to board members' investments in fossil fuels, and dyeing the water of Rome's famous Trevi Fountain black (Extinction Rebellion 2023; Arts 2023).

Despite these efforts, changemakers lack the political will necessary to implement the necessary policy change to prevent the worst scenarios of the climate crisis.

Gen Z is also employing protest as a strategy, nationally and internationally. In 2019, young protestors drew national attention after they confronted Democratic Senator Dianne Feinstein at her San Francisco office. The activists were from two climate activist groups: Youth vs. Apocalypse and the Sunrise Movement (The Atlantic 2019). Some have patronizingly criticized the young activists as "jackbooted tots and aggrieved teenagers" (Flanagan 2019), but members of the group defended their actions, describing how their intention was not to insult Feinstein but rather advocate for themselves and the issues they care about. As they explained, they respected Senator Feinstein's legislative experience and accomplishments, but they still reserved the right to question some of her policies as they would also be impacted by her decisions (Beckett 2019). In a video of the interaction between Senator Feinstein and the activists, Feinstein defended herself saying she was elected and earned her spot and that the children, who were not old enough to vote, had not voted for her. One ten-year-old protestor responded, "It doesn't matter. We're the ones who are going to be impacted" (Beckett 2019).

Internationally, Greta Thunberg, one of the most well-known Gen Z climate activists, started her efforts as a one-person school strike protesting outside of the Swedish parliament every Friday. Although she began her protest alone, her influence quickly spread as photos of her solo protest outside the Swedish parliament circulated, inspiring Fridays for Future, also known as School Strike for Climate. This movement is an international youth-led movement that encourages a reduction in global carbon emissions (Ottesen 2021). In honor of her bold actions, Thunberg was named *Time* magazine's "Person of the Year" in 2019 and has been nominated for the Nobel Peace Prize.

In a similar vein, activists across Europe have begun to draw attention to the climate crisis by symbolically defacing famous and treasured art such as paintings by Vincent Van Gogh and Claude Monet. Whether it is throwing mashed potatoes, tomato soup, or red paint, or gluing their hands onto the famous art, these Gen Zers bring dramatic awareness to the clear threat climate change poses. In Germany, the organization Last Generation that supported one such protest asked, "What is worth more, art or life?" (Medina 2022). Their efforts are particularly symbolic and poignant as the paintings they choose to deface are of beautiful landscapes, which they claim will no longer exist as the planet warms. (It should be noted that no art was actually damaged as they are protected by glass casings.)

As previously described, Gen Z is fighting at every level to affect change and slow the warming of the planet. Despite their relatively young age, their combined efforts are making a difference in personal choices, corporate decisions, and shaping the conversation in new and important ways with youth now being included in international-level policy talks. In general, Gen Z is motivated to work at the individual, local, national, and international levels to address a crisis, and their activism is making a difference.

Cultural change and popular will

Beyond the political arena, Gen Z is also shaping cultural values through the use of social media. This cohort, which has also been aptly named the "Igen," was exposed to technological advancements such as cell phones and the Internet since birth. Having grown up as digital natives, Gen Z is influencing the norms, beliefs, and behaviors through messaging on phone apps downloaded on billions of phones (Instagram, TikTok, Reddit, Pinterest, Facebook, etc.). For example, Lauren Ferree, a content creator on TikTok, posts videos on pressing environmental issues, spreading relevant information to over 225,000 followers on the platform. These types of content creators are increasingly shaping the views and behaviors of young adult audiences. Even one's choice of web browser can have an impact on the environment. For example, Google is a carbon-neutral platform (Calma 2020), and Ecosia is an alternative browser that commits to planting trees with every search (Ecosia. org 2024).

In this manner, Gen Z has the power to set trends. Indeed, in a Pew Research Center study, over half of Gen Z adults said that they had seen social media content calling for climate action in the weeks leading up to the survey, and nearly half said that they engaged with this content. Those who are engaging with the climate crisis on social media share different opinions than those who are not. For example, 87% of adults who engaged with climate content believed climate change is a top priority, but only 57% of those who do not engage believed the same (Funk 2021). An example of social media being used as a catalyst for environmental movements is the #StopWillow hashtag spreading through TikTok. The ConocoPhillips Willow Project would allow for oil drilling on the North Slope of Alaska. TikTok users have taken to using #StopWillowProject on videos and in comments to draw attention to the movement. As of March 2023, the hashtag had 66.9 million views (Nilsen 2023). Despite these efforts, President Joe Biden approved the project the same month. In response, Earthjustice filed a lawsuit opposing drilling on behalf of environmental activists, and the fight continues (Marris 2023).

In their private lives, some members of Gen Z are responding to the climate crisis by choosing to have fewer children. This is reflective in reproduction rates: the number of births has been in decline for the past two decades in the United States (Hamilton, Martin, and Osterman 2023) and in the Global North in general. While the reasons behind this trend are multifaceted, studies have found that Gen Z listed their worries about climate change as one of the top two reasons for not having children (Heins 2021). Gen Z's experiences with global warming are not only informing their attitudes but also shaping their behavior.

Expanding Generation Z's people power

If politicians and previous generations continue on the established path, future generations will face even more dire consequences of climate change. Since birth, Gen Z has learned about the causes of the climate crisis and experienced its dangers. In their lifetime, Gen Zers have witnessed some of the hottest temperatures and the deadliest climate change–related natural disasters on record, all while politicians do little to mitigate the underlying causes of the crises.

One of the greatest changes people can make to motivate engagement to fight against global warming is culturally reframing the climate crisis as a challenge that humanity must overcome rather than a foregone apocalyptic vision. This is particularly important for young people. Those members of Gen Z who are paralyzed by apathy or hopelessness must be "called in" (Bennett 2021) to replace doom scrolling, the excessive use of screen time devoted to reading dystopian news, to instead focus on envisioning environmentally sustainable strategies that are personal and feasible.

Solving the climate crisis is not solely an individual fight or a corporate, national, or international concern; no one is immune to changing weather patterns. However, Gen Z young adults, especially those who identify as middle and upper class, have power, privilege, and opportunity to take responsibility to be the vanguard of a sustainable society. As described earlier, this takes individual, local, national, and international changes addressing cultural, social, economic, and political structures that are slow to transform.

Our call

Gen Z activists are modeling to their apathetic peers and to older generations what climate activism looks like. We call in our generation and our elders to join us in demanding political and cultural change. While a lack of hope may paralyze, we cannot let it persuade us to ignore the issue or accept the status quo. Instead, we must follow the wisdom of Greta Thunberg who tweeted, "When we start to act hope is everywhere. So instead of looking for hope – look for action. Then the hope will come" (Thunberg 2018). We have the power to act by reflecting on and shifting our own behaviors while motivating others to do the same, challenging the societal status quo, demanding political action and corporate accountability, and envisioning a sustainable world – and we insist previous generations do the same!

References

Alfaro, Mariana. 2022. "Shaped by Gun Violence and Climate Change, Gen Z Weighs Whether to Vote." *The Washington Post*. Accessed March 7, 2023. www.washingtonpost.com/politics/2022/10/10/gen-z-voters-midterm-elections/

The Atlantic. "Members of Youth vs. Apocalypse Defend Their Exchange with Dianne Feinstein." *The Atlantic*. Accessed March 1, 2019. www.theatlantic.com/letters/archive/2019/03/youth-vs-apocalypse-respond-feinstein-and-green-new-deal/583852/

Arts, Avishay. 2023. "How Radical Should You be When You're Trying to Save the Planet?" *Vox*. Accessed October 28, 2023. www.vox.com/23892818/climate-change-activism-radical-protest-civil-disobedience

Baker, Mike. 2023. "A Landmark Youth Climate Trial Begins in Montana." *The New York Times*. Accessed November 20, 2023. www.nytimes.com/2023/06/12/us/montana-youth-climate-trial.html

Beckett, Lois. 2019. "You Didn't Vote for Me: Senator Dianne Feinstein Responds to Young Green Activists." *The Guardian*. Accessed November 9, 2023. www.theguardian.com/us-news/2019/feb/22/dianne-feinstein-sunrise-movement-green-new-deal

Bennett, Jessica. 2021. "What if Instead of Calling People Out, We Called Them in?" *The New York Times*. Accessed October 28, 2023. www.nytimes.com/2020/11/19/style/loretta-ross-smith-college-cancel-culture.html

British Broadcasting Company. 2020. "US West Coast Fires: I Don't Think Science Knows About Climate, says Trump." *BBC News*. Accessed November 20, 2023. www.bbc.com/news/world-us-canada-54144651

Bruggers, James, Darreonna Davis and Delaney Dryfoos. 2022. "The Nation's Youngest Voters Put Their Stamp on the Midterms, with Climate Change Top of Mind." *Inside Climate News*. Accessed March 7, 2023. https://insideclimatenews.org/news/10112022/young-voters-gen-z-climate-change/

Calma, Justine. "Google Announced One of the Biggest Green Pledges from Tech Yet." *The Verge*. September 14, 2020. Accessed November 20, 2023. www.theverge.com/2020/9/14/21436228/google-climate-change-pledge-2030-renewable-energy

Carnegie, Megan. "Gen Z: How Young People Are Changing Activism." *British Broadcasting Company*. Accessed November 20, 2023. www.bbc.com/worklife/article/20220803-gen-z-how-young-people-are-changing-activism

Copeny, Mary. 2023. "About Mari." *Mari Copeny*. Accessed June 24, 2023. www.maricopeny.com/about

Earthday.org. 2024. "Our History." Accessed March 4, 2024. https://www.earthday.org/history/

Earth Institute. 2010. "2000–2009: The Warmest Decade." *Columbia Climate School: The Earth Institute*. Accessed March 6, 2023. www.earth.columbia.edu/articles/view/2620

Ecosia.org. 2024. "The Search Engine that Plants Trees." Accessed March 4, 2024. https://www.ecosia.org/

Extinction Rebellion. 2023. "This Is an Emergency." Accessed October 28, 2023. https://rebellion.global/

Flanagan, Caitlin. "Dianne Feinstein Doesn't Need a Do-Over." *The Atlantic*. Accessed November 20, 2023. www.theatlantic.com/ideas/archive/2019/02/dianne-feinstein-video-climate-change-sunrise-movement/583501/

Funk, Cary. 2021. "Key Findings: How Americans' Attitudes About Climate Change Differ by Generation, Party and Other Factors." *Pew Research Center*. Accessed March 7, 2023. www.pewresearch.org/fact-tank/2021/05/26/key-findings-how-americans-attitudes-about-climate-change-differ-by-generation-party-and-other-factors/

Gammon, Katharine. 2022. "US Is Recycling just 5% of Its Plastic Waste, Studies Show." Accessed October 28, 2023. www.theguardian.com/us-news/2022/may/04/us-recycling-plastic-waste

Hamilton, Brady E., Joyce A. Martin and Michelle J. K. Osterman. 2023. "Births: Provisional Date for 2022." *Center for Disease Control and Prevention*. Accessed October 28, 2023. www.cdc.gov/nchs/data/vsrr/vsrr028.pdf

Heins, Camelia. 2021. "The Childless Generation: The Consequences of Opting Out of Having Children." *New University*. Accessed October 28, 2023. https://newuniversity.org/2021/12/14/the-childless-generation-the-consequences-of-opting-out-of-having-children/

Horvath, Hanna and Ambar Pardilla. 2021. "Climate Neutral: How to Shop for the Best Eco-Friendly Products." Accessed March 4, 2024. https://www.nbcnews.com/select/shopping/climate-neutral-how-shop-certified-eco-friendly-products-ncna1190891

Jamieson, Amie. 2002. "Voting and Registration in the Election of November 2000." *US Census Bureau*. Accessed November 9, 2023. www.census.gov/content/dam/Census/library/publications/2002/demo/p20-542.pdf

Kaufman, Mark. 2023. "The Carbon Footprint Sham." Accessed November 9, 2023. https://mashable.com/feature/carbon-footprint-pr-campaign-sham

Larson, Nina. 2020. "Climate Change Spurs Doubling of Disasters Since 2000: UN." *Phys. org.* Accessed March 6, 2023. https://phys.org/news/2020-10-climate-spurs-disasters.html#:~:text=Climate%20change%20is%20largely%20to,United%20Nations%20said%20on%20Monday

Lopez, Ashley. 2022. "Turnout Among Young Voters Was the Second Highest for a Midterm in Past 30 Years." *National Public Radio.* Accessed March 7, 2023. www.npr.org/2022/11/10/1135810302/turnout-among-young-voters-was-the-second-highest-for-a-midterm-in-past-30-years

Mackey, Robert. 2022. "Idaho's Far Right Suffers Election Loss to 18-year-old Climate Activist." *The Intercept.* Accessed November 9, 2023. https://theintercept.com/2022/09/13/idaho-boise-school-board-election/

Marris, Emma. 2023. "The Alaska Oil Project Will Be Obsolete Before It's Finished." *The Atlantic.* Accessed November 9, 2023. https://web.archive.org/web/20230313195458/www.theatlantic.com/science/archive/2023/03/biden-willow-alaska-arctic-oil-drilling/673382/

Medina, Eduardo. 2022. "Climate Activists Throw Mashed Potatoes on Monet Painting." *The New York Times.* Accessed October 28, 2023. www.nytimes.com/2022/10/23/arts/claude-monet-mashed-potatoes-climate-activists.html

Nature Conservancy. 2023. "What Is Your Carbon Footprint?" Accessed June 8, 2023. www.nature.org/en-us/get-involved/how-to-help/carbon-footprint-calculator/

Nilsen, Ella. "#StopWillow Is Taking TikTok by Storm. Can It Actually Work?" *CNN.* Accessed November 20, 2023. www.cnn.com/2023/03/05/us/willow-project-tiktok-petition-movement-climate/index.html

Ottesen, K. K. "Greta Thunberg on the State of the Climate Movement and the Roots of Her Power as an Activist." *The Washington Post.* Accessed November 20, 2023. www.washingtonpost.com/magazine/2021/12/27/greta-thunberg-state-climate-movement-roots-her-power-an-activist/

Parker, Laura. 2019. "Kids Suing Governments About the Climate: It's a Global Trend." *National Geographic.* Accessed October 28, 2023. www.nationalgeographic.com/environment/article/kids-suing-governments-about-climate-growing-trend

Ray, Sarah Jaquette. 2020. *A Field Guide to Climate Anxiety: How to Keep Your Cool on a Warming Planet.* Berkeley, CA: University of California Press.

Selin, Henrik and Stacy D. VanDeveer. 2009. "Local Government Responses to Climate Change: Our Last, Best Hope?" Essay. In *Changing Climates in North American Politics: Institutions, Policymaking, and Multilevel Governance*, 155. Cambridge, MA: MIT Press.

Thunberg, Greta. (@GretaThunberg), 2018. Twitter, October 21. Accessed November 20, 2023. https://twitter.com/GretaThunberg/status/1054048784844505098

Vidal, John. 2021. "The Environment in the Decade of Climate Change." *The Guardian*, Accessed October 28, 2023. www.theguardian.com/environment/2009/oct/17/environment-decade-climate-change-vidal

YouGov. 2022. "The Economy | Economist/YouGov Poll: August 7–9, 2022." *YouGov.* Accessed October 28, 2023. https://today.yougov.com/topics/economy/articles-reports/2022/08/10/economy-economist-yougov-poll-august-7–9–2022

6

QUEER POLITICAL CULTURE IN THE FACE OF THE CLIMATE CRISIS

Melanie M. Bowers and Cameron T. Whitley

The problem: Despite having many LGBTQ+ (lesbian, gay, bisexual, transgender, and queer) individuals actively working in the climate change movement, traditional climate change organizations and environmental decision-makers have largely ignored the concerns and interests of LGBTQ+ populations in their efforts to understand and prepare for climate change.

The solution: There is a growing LGBTQ+ climate change movement that functions both within and apart from the larger climate change movement. The activists and organizations working in this space engage in work to both respond to climate change and strengthen the LGBTQ+ community, often with the goal of improving LGBTQ+ climate resilience.

Where this has worked: LGBTQ+ climate organizations are approaching climate change work from a systems approach that does not decouple the environment from the individual, their intersecting identities, and their complex personal and societal needs. In this they can serve as a model for the broader climate change movement.

In a recent interview, queer climate activist Isaias Hernandez explained that "tackling the climate crisis is 'part of LGBTQ+ liberation . . . The idea of "queering" a system doesn't just come from putting a rainbow on it, it's actually putting justice on it'" (Chudy 2023). This sentiment encapsulates the spirit behind the growing queer climate movement, which is both part of and distinct from the broader climate change movement. LGBTQ+ people have always been an essential part of the

DOI: 10.4324/9781003437345-9

environmental movement, but in this chapter our focus is not on their participation in the broader environmental movement and traditional organizations. Instead, we explore a model for climate organizing: a decidedly queer approach to climate activism that is part of a broader queer environmentalism others have termed the eco-queer movement (see e.g., Sbicca 2012). The queer climate movement is successful in that it is intersectional, conceptualizes climate action through a systems lens, emphasizes the importance of community building, and largely, though not exclusively, functions through hyper-local organizations that emphasize micro actions.

The particular climate vulnerability of queer people

LGBTQ+ people's interest in climate change is often framed around the actual and potential harm the community faces from the climate crisis due to social and political marginalization. Indeed, climate change most dramatically impacts poor and socially disadvantaged people (Thomas et al. 2019), which in many places includes LGBTQ+ individuals (Dietz, Shwom, and Whitley 2020; Whitley and Bowers 2023). Even in the United States where the community has relatively high social and political standing, LGBTQ+ people, especially transgender people and people of color, remain marginalized and tend to have lower incomes (Badgett et al. 2019), less housing stability (Conron 2019; Conron et al. 2018), higher rates of homelessness (Ecker, Aubry, and Sylvestre 2019; Fraser et al. 2019), more family and social violence (Adamson et al. 2022; Turner and Hammersjö 2023), higher rates of mental health problems and addiction (Clark et al. 2020; Moazen-Zadeh et al. 2019), and less access to healthcare (Dawson, Long, and Frederikson 2023). These conditions are likely to be amplified by the types of food and supply shortages, structural damage, and social unrest predicted by a changing climate (Goldsmith, Raditz, and Méndez 2021; Whitley and Bowers 2023).

Natural disasters have highlighted LGBTQ+ individuals' precarious positions in times of emergency. When communities are devastated by climate-induced disasters, their residents must seek shelter, food, healthcare, and general assistance from the community at large. Because anti-LGBTQ+ sentiments and discrimination are prominent throughout the world, this has meant that LGBTQ+ individuals face myriad problems during natural disasters that range from being unable to access shelter to physical violence (Dominey-Howes, Gorman-Murray, and McKinnon 2014; Yamashita et al. 2017).

While it is essential that we understand the direct impacts that LGBTQ+ people face due to climate change – and doing so provides one explanation for why queer activists talk about climate change as an LGBTQ+ issue – focusing on harm alone suggests that the LGBTQ+ community's relevance to the conversation largely stems from their vulnerability and their need to be protected. There is a paternalism in this approach that ignores the community's demonstrated capacity to promote change and the depth of organizing experience, innovative thinking, and creative problem-solving that LGBTQ+ people bring to the table.

Activists and organizations working in the queer climate movement are acutely aware of the community's power and potential to instigate change. Here, we start from the premise that the LGBTQ+ community has a unique political culture that primes it for activism and makes it a particularly powerful agent of change.

The power of queer culture and environmental activism

In the fall of 2023, LGBTQ+ climate activist Jerome Foster II made headlines when he was appointed to be a climate change advisor to the Biden-Harris administration, becoming the youngest White House advisor in history. " 'Activists are instruments of disruption in any space that we're in,' he said. 'The most powerful thing we can do is to shake up the system anywhere we can'" (Benn 2023). The 21-year-old embodies much of what we see in the queer climate movement: he is young, approaches the work from an intersectional lens, and is focused on the dual goal of environmental protection and LGBTQ+ rights. His decision to reject an invitation to the 27th Conference of the Parties to the United Nations Framework Convention on Climate Change because it was held in Egypt, where queer people are brutally persecuted, emphasizes a common view in the movement: the fate of LGBTQ+ people and the environment are inextricably connected; protection of one is not possible without protection of the other. That there is a uniquely queer submovement within the broader environmental movement is not particularly surprising. LGBTQ+ people are uniquely positioned to be activists and promote climate action and environmental change but do so through an experience of exclusion that also encourages community protection.

We are in the early days of organizing LGBTQ+ people around climate change and climate-induced disasters, but the community's long history with organizing and activism has imbued it with human capital and institutional knowledge that contributes to a broader culture of activism. Stonewall, the AIDS crisis, and the fight for civil rights and social acceptance have created a communal understanding that political cohesion is necessary for social and political change.

Through a history of economic, social, and political exclusion, the LGBTQ+ community has developed a political and social justice foundation that, while originating from the need to survive, has evolved into a communal sense of political efficacy that suggests the world can be changed through collective action. This idea is captured in the classic queer political chant "not gay as in happy, but queer as in fuck you!" We and other scholars have argued that this represents a distinct queer political culture. This sense of cohesion and extensive in-group networks poise the community to mobilize around political and environmental issues. As the Institute for Queer Ecology explains,

Queer communities are uniquely positioned to lead on climate adaptation through embodied strategies already inherent or familiar to queer experiences. On an individual level, queer lives are mutable: we understand change and transformation . . . On a collective level, queer community is mutualistic: it is symbiotic, in-contact, relational . . . (Institute for Queer Ecology 2023)

Queer political culture helps explain why LGBTQ+ people appear to be particularly attuned to the climate crisis and hold distinct views on climate change. In a recent survey we conducted, LGBTQ+ individuals expressed higher agreement with belief in climate change, identified climate change as a greater threat, and expressed more worry about climate change than their cisgender, heterosexual (cishet) counterparts (Whitley and Bowers 2023). These findings mimic earlier work showing that higher portions of the LGBTQ+ population reported caring about green issues and being concerned with politicians' environmental views than their cishet counterparts (Harris Interactive 2010). This highlights the importance the LGBTQ+ population places on environmental issues in the United States and, when combined with a broader understanding of queer political culture, helps us understand the queer climate movement.

To summarize, a history of oppression and activism has created a multifaceted political culture that encourages LGBTQ+ people to gain tangible skills that can be used for activism on virtually any issue, facilitates a sense of collective identity that is both separate from society at large and oriented toward societal transformation, and, because of a legacy of successful organizing and political change, encourages a sense of efficacy that inspires additional action. Because LGBTQ+ people are also more concerned about the environment and climate change on average, queer political culture creates a natural path to develop a queer climate movement that simultaneously emphasizes environmental and LGBTQ+ protection.

General trends in the queer climate movement

The environmental movement has been critiqued for being largely White, middle class and heteronormative (see e.g., Bullard 2015; Ghoche and Udoh 2023; Sandler and Pezzullo 2007). Many environmental organizations and activists have acknowledged and responded to these critiques in meaningful ways. For example, recognizing that representation is important, between 2018 and 2022 the world's largest environmental nongovernmental organizations (NGOs) increased the percentage of their leadership and staff identifying as people of color from around 20% (Johnson 2019) to 31% of leaders and nearly 40% of staff (Green2.0 2022). Despite these advancements, the movement has yet to embrace the truly intersectional approach that scholars like David N. Pellow (2018) – see Chapter 1 in this volume – and Robert Bullard (2015) argue are necessary for transformation.

Pellow, Bullard, and other critical environmental justice scholars tell us that the environmental movement must do the following:

1) Take an intersectional approach that acknowledges multiple sources of oppression and inequality across varying spatial scales.
2) Understand and acknowledge that the nation-state creates the very inequalities that make environmental harm so catastrophic and therefore cannot be the only or even primary source of redress.
3) Center and emphasize the experiences of those communities that are most harmed by environmental problems.

For a relatively technocratic field that has emphasized formal state action as a primary mechanism for change, this is a tall order. Yet, in this, the queer climate movement stands apart, emphasizing intersectionality, the intentional centering of those most greatly affected by climate change, and largely, though not exclusively, hyper-local and extra-governmental action.

Perhaps the most defining characteristic of the queer climate movement is its intersectional nature (Gaard 2019; Garwood 2016; Pakin-Albayrakoğlu 2022). These organizations typically start from the position that individuals experience the consequences of climate change differently depending on their intersecting identities and the inequality and oppression they already face. This means expressly acknowledging the distinct experiences of transgender people, communities of color, and trans people of color and working to center the experiences of those most negatively impacted by climate change. For example, Our Climate Voices is a youth-led organization that elevates the experiences of those most affected by climate change through storytelling. Among other things, the organization has artists in residence who engage in workshops around storytelling and expressing the experiences of those affected by climate change, hosts trainings on ethical climate change storytelling, and provides mechanisms for the broader community to financially support storytellers, elevating the voices of people of color, Indigenous communities, women, LGBTQ+ people, and others most greatly affected by climate change (Our Climate Voices 2023).

As part of the intersectional approach, there is an awareness that efforts to overcome oppression are interconnected and cannot be successful in isolation. Consequently, this movement and the activists who work within it often recognize liberation movements working for nonenvironmental economic, racial, immigrant (etc.) justice as being linked to the success of the environmental movement. The implications of this belief vary by organization but are incorporated into mission statements, shape the work that is undertaken, and are reflected in formal statements of support, coalition formation, and allyship. For example, for Seattle-based Out for Sustainability's mission, two of its four pillars recognize an alignment with other liberation or justice movements. The first pillar, redistribution, recognizes that food, energy, and water security are necessary for climate risk management;

the second, transformation, expressly states the organization "aligns with movements for queer and transgender liberation and ecological justice, including calls for climate reparations, land repatriation, food sovereignty, and bodily autonomy" (Out for Sustainability 2023). To fulfill this and other parts of its mission, the organization engages in grassroots organizing, mutual aid, fiscal sponsorship for other climate organizations, microlending, and education.

Importantly, at least in part, the view of interconnected injustice stems from the intentionally queer nature of the movement. The movement is not simply LGBTQ+ people working in environmental organizations; it is unapologetically queer *and* environmental, meaning that while the organizations and activists involved focus on climate action, they also emphasize the growth and support of the queer community. This manifests as everything from the Institute of Queer Ecology curating art by queer artists that emphasizes ecological themes (Institute for Queer Ecology 2023) to Queer Nature's workshops, which train LGBTQ+ people in wildlife survival skills to encourage resilience (Queer Nature 2023). Organizations thus focus on environmental problems but do so while also working to promote the interests of the LGBTQ+ community. This means that queer climate organizations' work often includes things like emphasizing LGBTQ+ inclusion in the broader climate movement, increasing LGBTQ+ representation and concerns in conventional environmental decision-making, improving LGBTQ+ populations' climate change resilience, and working to build LGBTQ+ community strength above and beyond its environmental resilience. In its "Who We Are" statement, Queers X Climate explains, "In the case of climate change, this is arguably the single most dangerous threat ever faced by humanity. All the development accomplishments, including the achievements on LGBTQ+ rights, could be erased within barely a decade by exacerbating resource scarcity and social unrest" (Queers X Climate 2023). There is thus a self-conscious understanding that the fate of LGBTQ+ people is inextricably connected to a peaceful, stable world and political system that are dependent on a healthy environment and climate.

Because the queer climate movement has the dual goal of promoting environmental change and supporting the LGBTQ+ community, its methods are diverse, reflect a systems approach, and tend toward smaller-scale, community-based action. Systems theory is used across disciplines to evaluate phenomena holistically, looking for the interconnectedness of systems, beliefs, causal mechanisms, and outcomes. Building off its foundational understanding of intersectionality and interconnected vulnerability, many organizations within the queer climate movement approach climate action from a systems lens that views broader economic, housing, and food security as well as community strength as being linked to climate resilience. Consequently, while some queer climate organizations engage in work that is common to the broader climate movement, such as providing education on climate change, many are focused on things like creating environmental engagement opportunities for LGBTQ+ people or promoting food security and sustainable agriculture while building LGBTQ+ community. For those outside of the movement, this type of work may not immediately be recognized as climate action, since it often lacks a direct tie to things like greenhouse gas emissions, but for the organizations engaged in this work, encouraging environmental connection, food and economic security, and ecological knowledge in the LGBTQ+ community is viewed as essential to prepare the community for climate change.

This type of hyper-local action harkens back to the lesbian-led back to the land movement of the 1960s, where female separatists developed rural communities promoting sustainable models of living off the land that emphasized symbiosis with nature (Unger 2012). POC-led Hilltop Urban Gardens (HUGs) offers a prime example of how webbed the queer climate movement often views climate change and other social issues. At a foundational level, HUGs links climate adaptation to the BIPOC community's general ability to thrive and thus conceptualizes its work on decolonization and land and food sovereignty as climate change activism. As part of the organization's work to respond to climate change, it is developing a wellness and ecology park on vacant land in Tacoma, Washington, that will, among other things, include medicinal gardens. The herbs grown in the gardens will be used in HUGs Black birthworker initiative, where queer/trans Black birthworkers will be trained to work in herbal clinics spread throughout King County with the goal of reducing Black maternal and infant mortality (Hilltop Urban Gardens 2023). This type of intersectional, interconnected, hyper-local work captures significant aspects of the queer climate movement, which pushes the boundaries of what we think of as climate change work.

Capacity for change

There are at least two key questions to ask about any social movement: Does the movement have the capacity to be successful and achieve its goals? Has it been successful in producing change? The queer climate movement's grassroots nature, nonviolent approach, consistent messaging, and persistence provide tics in the "successful movement" column, but its decentralized nature, varied goals, and limited alliance building create limitations for its reach. We talked about the queer climate movement's central tendencies: it is intersectional, interconnected, focused on the environment and queer community, and tends to be hyper-local. These commonalities mean that the movement has a strong grassroots orientation that emphasizes the importance of utilizing existing skills while building community capacity, things that are prerequisites to successful social movements (see e.g., Jenkins and Halcli 1999). The movement also has consistent messaging across its organizations and activists that frames climate change as an LGBTQ+ issue and links LGBTQ+ rights and climate justice to other social and economic problems and movements. This type of clear messaging helps attract supporters and communicates the importance of action (see e.g., Snow and Benford 1988).

At the same time, the movement's tendency toward hyper-local responses also means it is decentralized, without a clear leader and with underdeveloped alliance formation, traits that make change more difficult (see e.g., Van Dyke and McCammon 2010). Though most queer climate organizations recognize the importance of allyship, building alliances across issue areas, and showing up for other movements, the individual organizations tend to be small and geographically separated, with limited capacity. As a result, we see limited evidence of coalition building across organizations and activists within the movement, for example, in the form of joint campaigns or projects, shared visioning, collective education efforts, collectively organized rallies, or pressure campaigns. The notable exception is that a variety of organizations and activists advertise LGBTQ+ climate events organized throughout the world by other groups.

That said, beyond coalition building with like-minded organizations, we know that successful movements do what Schattschneider (1960) described as "expanding the scope of conflict," meaning that they engage groups and individuals not directly tied to the movement to fight on their behalf. For the queer climate movement, this means finding groups who will fight specifically for the needs of LGBTQ+ people. In this way, the movement is making some inroads. In recent years, we have seen traditional environmental organizations like Greenpeace increasingly discussing climate change impacts on the LGBTQ+ community, likely because of the pressure, educational efforts, publications, and awareness campaigns that LGBTQ+ climate activists have created (see e.g., Premkumar and Atanasova 2023). Further, some LGBTQ+ climate organizations are developing relationships with government entities to improve LGBTQ+ outcomes. For example, Out for Sustainability recently produced a report about how to incorporate the needs of LGBTQ+ populations in disaster planning based on webinars it hosted with the Federal Emergency Management Agency and the US Department of Homeland Security (Goldsmith 2023). Although the movement tends toward extra-governmental action, this type of work and efforts to increase LGBTQ+ representation in environmental decision-making will likely be necessary to achieve the types of consideration and policy change needed for LGBTQ+ people to have their climate change needs and concerns addressed.

When it comes to the question of success, the answer is multifaceted. Most queer climate organizations have a goal to improve local LGBTQ+ climate resilience. This has been accomplished through training LGBTQ+ people in survival and agricultural skills, advocating for LGBTQ+ needs in disaster planning, and engaging in educational efforts within the LGBTQ+ community. When it comes to broader impacts, these organizations do not intend to have widespread impacts on climate change itself (e.g., by reducing emissions) and are not likely to produce them.

The fact that the movement focuses on small-scale actions, however, does not mean that it is only influential on a local level. One way that queer climate activists have the potential for greater impact is by pushing the larger climate change movement to embrace intersectionality, improve its reach and diversity, and embrace out-of-the-box ideas for change. For example, Greta Gaard describes how 11 LGBTQ+

employees at 350.org pushed the environmental movement to embrace the concerns of the LGBTQ+ community after the Pulse Nightclub shooting,[1] explaining:

> The internationally known climate justice organization, 350.org sent out a collectively authored message of grief and hope affirming "our fights are connected," and "as LGBTQ+ climate activists, we need to bring our whole selves to this work." Disclosing that "many of us who are shoulder to shoulder with you in the streets are LGBTQ+."
>
> *(Gaard 2019:92)*

We have also seen conventional organizations increasingly centering the needs and experiences of people of color and Indigenous communities that are most significantly impacted by climate change in ways that reflect an intersectional understanding. While the queer climate movement cannot take credit for this change, which reflects significant effort on the part of communities of color, activists and organizations as well as changing cultural norms and expectations, it is part of the general effort to push the climate movement toward intersectionality. As it gains momentum, the movement also has the potential to push climate organizations to expand their conception of climate action in ways that increase their reach and audience.

In summary, the calls in this volume and elsewhere for democratic, community-driven organizing that takes into account the lived experiences of marginal people are modeled in the existing queer climate movement. This movement serves as a model for what successful, grassroots, justice-oriented climate organizing looks like.

Note

1 The Pulse Nightclub shooting was one of the most deadly attacks against the LGBTQ+ community in the United States. In 2016 Omar Mateen entered the Pulse Nightclub, an LGBTQ+ nightclub in Orlando, Florida, and murdered 49 people, injuring 53 others.

References

Adamson, Tyler, Elle Lett, Jennifer Glick, Henri M. Garrison-Desany and Arjee Restar. 2022. "Experiences of Violence and Discrimination Among LGBTQ+ Individuals During the COVID-19 Pandemic: A Global Cross-Sectional Analysis." *BMJ Global Health* 7(9): e009400.

Badgett, M. V. Lee, Soon Kyu Choi and B. D. Wilson. 2019. *LGBT Poverty in the United States*. Los Angeles, CA: The Williams Institute and American Foundation for Suicide.

Benn, Cal. "White House Climate Change Advisor Makes History." *Washington Blade*. Retrieved November 1, 2023. www.washingtonblade.com/2023/10/27/white-house-climate-change-advisor-makes-history/#:~:text=A%2021%2Dyear%2Dold%20LGBTQ,House%20Environmental%20Justice%20Advisory%20Council

Bullard, Robert D. 2015. "Environmental Racism and the Environmental Justice Movement." In *Thinking About the Environment*, 196–204. New York: Routledge.

Chudy, Emily. 2023. "Queer Environmental Activists Warn of LGBTQ+ People 'Suffering First' from Catastrophic Climate Change." *The Pink News.* Retrieved November 1, 2023. www.thepinknews.com/2023/08/24/climate-change-queer-activists-lgbtq-environmental-crisis/

Clark, Kirsty A., Susan D. Cochran, Anthony J. Maiolatesi and John E. Pachankis. 2020. "Prevalence of Bullying Among Youth Classified as LGBTQ Who Died by Suicide as Reported in the National Violent Death Reporting System, 2003–2017." *JAMA Pediatrics* 174(12): 1211–1213.

Conron, Kerith. 2019. *Financial Services and the LGBTQ+ Community: A Review of Discrimination in Lending and Housing. Testimony Before the Subcommittee on Oversight and Investigations.* Williams Institute. Retrieved November 1, 2023. https://williamsinstitute.law.ucla.edu/wp-content/uploads/Housing-and-Lending-Testimony-Oct-2019.pdf

Conron, Kerith J., Shoshana K. Goldberg and Carolyn T. Halpern. 2018. "Sexual Orientation and Sex Differences in Socioeconomic Status: A Population-Based Investigation in the National Longitudinal Study of Adolescent to Adult Health." *Journal of Epidemiology and Community Health* 72(11): 1016–1026.

Dawson, Lindsey, Michelle Long and Brittni Frederiksen. 2023. "LGTB+ People's Health Status and Access to Care." *KFF.* Retrieved November 3, 2023. www.kff.org/report-section/lgbt-peoples-health-status-and-access-to-care-issue-brief/#:~:text=One%2Dquarter%20(25%25)%20of,by%20race%2Fethnicity%20or%20gender

Dietz, Thomas, Rachael L. Shwom and Cameron T. Whitley. 2020. "Climate Change and Society." *Annual Review of Sociology* 46: 135–158.

Dominey-Howes, Dale, Andrew Gorman-Murray and Scott McKinnon. 2014. "Queering Disasters: On the Need to Account for LGBTI Experiences in Natural Disaster Contexts." *Gender, Place & Culture* 21(7): 905–918.

Ecker, John, Tim Aubry and John Sylvestre. 2019. "A Review of the Literature on LGBTQ Adults Who Experience Homelessness." *Journal of Homosexuality* 66(3): 297–323.

Fraser, Brodie, Nevil Pierse, Elinor Chisholm, and Hera Cook. 2019. "LGBTIQ+ Homelessness: A Review of the Literature." *International Journal of Environmental Research and Public Health* 16(15): 2677–2690.

Gaard, Greta. 2019. "Out of the Closets and Into the Climate! Queer Feminist Climate Justice." In *Climate Futures: Reimagining Global Climate Justice,* edited by K. Bhavnani, J. Foran, P.A. Kurian and D. Munshi, 92–101. England: Zed Books, Ltd.

Garwood, Eliza. 2016. "Reproducing the Homonormative Family: Neoliberalism, Queer Theory and Same-Sex Reproductive Law." *Journal of International Women's Studies* 17(2): 5–17.

Ghoche, Ralph and Unyimeabasi Udoh. 2023. "A Commons for Whom? Racism and the Environmental Movement." In *Transforming Education for Sustainability: Discourses on Justice, Inclusion, and Authenticity,* 75–88. New York City: Springer.

Goldsmith, Leo. 2023. "Inclusive and Equitable Emergency Management." *Out for Sustainability.* Retrieved November 1, 2023. https://img1.wsimg.com/blobby/go/cd599488-9b66-4c7c-bfc2-68da8aef9363/Report%20Inclusive%20and%20Equitable%20Emergency%20Manag.pdf

Goldsmith, Leo, Vanessa Raditz and Michael Méndez. 2021. "Queer and Present Danger: Understanding the Disparate Impacts of Disasters on LGBTQ+ Communities." *Disasters.*

Green2.0.2022. "2022 NGO and Foundation Transparency Report Card." Retrieved November 5, 2023. https://diversegreen.org/transparency-cards/2022-green-2-0-ngo-foundation-report-card/

Harris Interactive. 2010. "Environmental Advocacy Grows Stronger for LGBT Americans." Accessed September 3, 2023. https://www.prnewswire.com/news-releases/environmental-advocacy-grows-stronger-forlgbt-americans-111941139.html

Hilltop Urban Gardens. 2023. "Projects and Initiatives." Retrieved November 5, 2023. www.hilltopurbangardens.org/services-1

Institute for Queer Ecology. 2023. "The Institute for Queer Ecology." Retrieved November 5, 2023. https://queerecology.org/About

Jenkins, J. Craig and Abigail Halcli. 1999. "Grassrooting the System? The Development and Impact of Social Movement Philanthropy, 1953–1990." In *Philanthropic Foundations: New Scholarship, New Possibilities*, edited by E.C. Lagemann, 229–256. Bloomington, IN: Indiana University Press.

Johnson, Stefanie K. 2019. "Leaking Talent: How People of Color Are Pushed out of Environmental Organizations." *Green2.0*. Retrieved November 5, 2023. https://diversegreen.org/wp-content/uploads/2021/01/Green_2.0_Retention_Report.pdf

Moazen-Zadeh, Ehsan, Mohammad Karamouzian, Hannah Kia, Travis Salway, and Olivier Ferlattand Rod Knight. 2019. "A Call for Action on Overdose Among LGBTQ People in North America." *The Lancet Psychiatry* 6(9): 725–726.

Our Climate Voices. 2023. "Our Climate Voices." Retrieved November 13, 2023. https://actionnetwork.org/groups/our-climate-voices

Out for Sustainability. 2023. "Mission & Strategy." Retrieved November 10, 2023. https://out4s.org/who-we-are

Pakin-Albayrakoğlu, Esra. 2022. "Out and Proud in the Field: Eco-Queers for Climate Adaptation." *Peace Review* 34(1): 51–63.

Pellow, David N. 2018. *What Is Critical Environmental Justice?* Cambridge, UK: Polity Press.

Premkumar, Shanthuru and Lina Atanasova. "The Disproportionate Impact of Climate Change on the LGBTQIA2S+ Community." *Greenpeace*. Retrieved November 10, 2023. www.greenpeace.org/international/story/60078/impact-climate-crisis-lgbtqia2spride-month

Queer Nature. 2023. "Our Mission." Retrieved November 10, 2023. www.queernature.org/what-we-do

Queers X Climate. 2023. "Who We Are." Retrieved November 10, 2023. www.queersxclimate.org/about

Sandler, Ronald D. and Phaedra C. Pezzullo. 2007. *Environmental Justice and Environmentalism: The Social Justice Challenge to the Environmental Movement*. Boston: MIT Press.

Sbicca, Joshua. 2012. "Eco-queer Movement (s)." *European Journal of Ecopsychology* 3: 33–52.

Schattschneider, Elmer Eric. 1960. *The Semisovereign People: A Realist's View of Democracy in America*. Boston: Cengage Learning.

Snow, David A. and Robert D. Benford. 1988. "Ideology, Frame Resonance, and Participant Mobilization." *International Social Movement Research* 1(1): 197–217.

Thomas, Kimberley, R. Dean Hardy, Heather Lazrus, Michael Mendez, Ben Orlove, Isabel Rivera-Collazo, J. Timmons Roberts, Marcy Rockman, Benjamin P. Warner and Robert Winthrop. 2019. "Explaining Differential Vulnerability to Climate Change: A Social Science Review." *Wiley Interdisciplinary Reviews: Climate Change* 10(2): e565.

Turner, Russell and Anjelica Hammersjö. 2023. "Navigating Survivorhood? Lived Experiences of Social Support-Seeking Among LGBTQ Survivors of Intimate Partner Violence." *Qualitative Social Work* 23(2): 242–260. https://doi.org/10.1177/14733250221150208.

Unger, Nancy C. 2012. *Beyond Nature's Housekeepers: American Women in Environmental History*. New York: Oxford University Press.

Van Dyke, Nella and Holly J. McCammon. 2010. *Strategic Alliances: Coalition Building and Social Movements*. Minneapolis, MN: University of Minnesota Press.

Whitley, Cameron T. and Melanie M. Bowers. 2023. "Queering Climate Change: Exploring the Influence of LGBTQ+ Identity on Climate Change Belief and Risk Perceptions." *Sociological Inquiry* 93(3): 413–439.

Yamashita, Azusa, Christopher Gomez and Kelly Dombroski. 2017. "Segregation, Exclusion and LGBT People in Disaster Impacted Areas: Experiences from the Higashinihon Dai Shinsai (Great East-Japan Disaster)." *Gender, Place & Culture* 24(1): 64–71.

PART 3

Changing the stories

7
OVERCOMING HURDLES TO CLIMATE MITIGATION

How motivational barriers impact strategies for change

Samantha Noll

The problem: The sheer scale of the harms produced by climate change makes it a central issue in the national and global arena. Yet, societies, as well as concerned citizens, are struggling to respond effectively. Adaptation and mitigation strategies are hindered by motivational barriers arising from the "wicked" nature of the problem. Even if we recognize a duty to mitigate impacts, there are social factors that hamper effective change at the policy and individual levels – factors that ultimately curtail the effectiveness of strategies to fight climate change.

The solution: The social sciences and humanities provide valuable insights concerning wicked problems. Wicked problems are challenges that are complex and require expertise from multiple disciplines to address. These disciplines help create strategies to address motivational barriers. Social science research informs contemporary mitigation strategies and has the potential to help communities work toward effectively addressing climate change.

Where this has worked: Strategies informed by the social sciences are being implemented with success especially at the regional policy level. For example, the US Department of Agriculture (USDA) has developed a plan to mitigate the effects of climate change on agriculture, forests, and rural communities around the country. One of their goals is to reduce the vulnerability of human and natural systems and exposure to effects. They hope to accomplish this by creating Climate Hubs, or regional offices, to develop locally specific tools and resources to empower farmers, ranchers, and land managers to manage harms on the ground.

DOI: 10.4324/9781003437345-11

The 21st century is the warmest period on record since humans started recording global temperatures in 1880 (Heeter et al. 2023). This makes the problem of climate change a central issue in the national and global arena. Today, many nations have taken significant actions to fight climate change, demonstrating their commitment to reduce emissions and adapt to changing temperatures (Basseches et al. 2022). In addition, prominent organizations, such as the United Nations, argue that individuals can also contribute to collective efforts. They recommend changing daily behavior to help fight the climate crisis, including saving energy at home, taking public transit, rethinking travel plans, etc. (United Nations 2021).

Yet, climate mitigation and adaptation strategies face policy gridlock, rollbacks, and other barriers to action at every turn (Basseches et al. 2022). For instance, at the individual level, even though we have clear recommendations and people understand that climate change is a threat (Saad 2013), individuals often do little to modify personal behavior, especially when they feel that their actions will not make a difference (Salomon, Preston, and Tannenbaum 2017). Similarly, while the scientific community began to call for action in the 1980s (Pester 2021), international treaties, policies, and regulatory responses have been slow to develop (Grundmann 2016).

I have studied environmental problems for over 10 years. During that time, I've learned a great deal about the factors impacting our ability to act when facing ecological challenges (Noll 2017, 2018a; Glazebrook, Noll, and Opoku 2020). I've also written extensively on how we can overcome barriers to climate adaptation strategies and practice (Noll 2017, 2018b, 2023). My work focuses on agriculture and food impacts, as these are important for food security and to me, personally. For instance, my family has been farming for generations. Shifting weather patterns could mean that rains fail to come when they are supposed to, killing crops from either a lack of water or too much water, causing good food to rot in the field. Similarly, unseasonal warm weather followed by an early frost can kill a tree fruit crop, as blooms freeze before they turn into apples, cherries, and other fruits that we enjoy. For many farming families, failed harvests mean unpaid bills, and looming bankruptcy. The US Department of Agriculture (USDA 2013) is aware of these impacts, as climate change is projected to have detrimental effects on *most* crops and livestock grown in the United States, and beyond. Agriculture is a hard business, in the best of times. It is easy to spot the marks that a changing climate leaves on the earth and in people's lives, especially if you know what you are looking for. Farmers throughout the country are already seeing the signs, yet we have been slow to respond to this looming threat.

As discussed in the next section, one reason for this slow response is the nature of climate change as a "wicked" problem (Duckett et al. 2016; Grundmann 2016). Wicked problems are highly complex, with numerous and sometimes undefined causes that have culturally contested understandings. Society struggles to respond effectively to wicked problems, so it is not surprising that mitigation efforts both in the United States and globally have been slow, uneven, and socially divisive (Basseches et al. 2022).

The chapter begins by discussing how climate change is now largely recognized as a threat to human life, yet attempts to mitigate impacts have met with limited success. It explores this bottleneck, highlighting how adaptation and mitigation strategies could be hindered by motivational barriers arising from the "wicked" nature of the problem. Even if societies and citizens recognize a duty to address climate impacts, there are obstacles that could ultimately curtail the effectiveness of potential strategies. These include, but are not limited to, goal barriers, ambiguity barriers, and threat barriers, as well as ethical challenges curtailing immediate action. The social sciences and humanities provide insights into the social aspects of wicked problems, as well as strategies to address each of these barriers. This research is important if we hope to effectively motivate citizens and society to take steps to mitigate climate impacts. The aim of this chapter is to introduce readers to this important area of research while also providing the foundation necessary for further work on motivational barriers and ethical challenges. Understanding social barriers to action is imperative for effectively addressing what has been called the most wicked problem of the modern age: climate change.

Climate change is a wicked problem

While changes to the climate are increasing (Tollefson 2021), attempts to mitigate impacts have met with limited success (Shalev 2015; Whitmarsh and O'Neill 2010). As Moser and Ekstrom (2010:22026) argue, "adaptation to climate change has risen sharply as a topic of scientific inquiry . . . Yet climatic events in Europe, the United States, and Australia in recent years have also led to critical questioning of richer nations' ability to adapt to climate change." A perfect storm of motivational barriers and ethical challenges combined to hamper societal and personal mitigation strategies. This is not surprising, as climate change is a "wicked problem," a multifaceted and systemic issue that has no simple solution (Whyte and Thompson 2011). Wicked problems are uniquely challenging, as even the way we conceptualize them is nebulous (is climate change an environmental, economic, or social problem?) and solutions are varied, depending on various contextual factors and how we frame the problem. Whyte and Thompson (2011:441–442) capture this complexity well in the following statement:

To describe climate change as an economic problem means that one has already limited oneself to particular economic solutions to addressing it. Because proposed solutions are so closely tied to problem formulations, disagreements among stakeholders who foresee themselves as being impacted differently by the solutions can take the form of ontological debates. Unlike problems where there is little disagreement about its basic formulation, wicked problems are characterized by deep ambiguity in the ontological assumptions and metaphysical categories used in their articulation.

Climate change is one of the most intricate challenges that we face, as it involves interactions between biological processes and a diverse array of human conduct and social systems (Noll 2023). To further complicate matters, this phenomenon has been labeled a "super wicked problem," due to four factors that greatly increases its overall complexity beyond typical challenges associated with these types of problems. According to Levin et al. (2012:123), these are the following: it is a time-sensitive issue; the problem is caused by those who also seek a solution; the governmental authority needed to enact change is weak or nonexistent; and as a result, policy responses discount the irrationality of future decisions. The authors go on to argue that these factors resulted in a policymaking "tragedy," where traditional political systems are ill equipped to identify and implement solutions. It follows from this categorization that there is no one clear policy solution or goal to guide individual behavior. As we discuss, both social and ethical expertise are necessary to remove motivational barriers hindering climate mitigation and adaptation strategies.

Motivational barriers and ethical challenges

Before we discuss strategies to address barriers, we must first define what they are. Due to the nature of climate change, local, national, and international discourses often focus on weighing the merits of various goals and solutions (Noll 2023; Shalev 2015). According to Shalev (2015), there is an abundance of social science research that connects the prolonged evaluation and assessment of goals to decision-making paralysis (Kruglanski et al. 2010; Shalev and Sulkowski 2009). If this is the case, then in addition to complexity hampering policy-focused solutions, climate change decision-making may also be hampered by motivational barriers. According to Shalev (2015), an important barrier to climate change adaptation is the lack of motivation, which is understood as the process that elicits individual people to act. Human motivation can be divided into two categories: motivations that do not involve conscious awareness and those that require conscious thought (Baumeister and Bargh 2014). The latter helps individuals plan for future contingencies and thus is important for climate change strategies. There are three types of barriers that impact conscious motivations: goal barriers, threat barriers, and ambiguity barriers (Shalev 2015). As we see, each type plays a role in reducing individual's efforts to help mitigate climate change.

The three types of barriers are relatively straightforward. Concerning the first type, clearly defined goals and endpoints are imperative for pursing solutions (Kruglanski et al. 2010; Noll 2023; Shah et al. 2002). A clear sense of direction is necessary for providing the personal impetus necessary to realize specific goals (Shalev 2015). When there are too many goals, or they are contested and/or ambiguous, individuals will fail to act quickly. Similarly, people react counterintuitively to threats, often focusing on achieving immediate goals, rather than trying novel solutions to nebulous threats. As Shalev (2015) argues, threats inhibit a person's ability to change behavior or try new experiences. Thus, "individuals who feel they are under threat . . . tend to neglect their long-term, future planning goals in favour of the short-term goal of self-defence" (Shalev 2015:131). Threat often motivates individuals to reaffirm familiar behavioral patterns rather than changing them (Steele 1988; Cohen, Janicki-Deverts, and Miller 2007). Paradoxically, this can reduce a person's ability to modify behavior, even when faced with new evidence that novel solutions are needed (Shalev 2015). Finally, ambiguity also increases resistance to behavioral change (Jost et al. 2003), as do situations where a rapid response is needed (Kruglanski 2004).

When taken as a whole, the wicked nature of climate change could potentially produce each of the motivational barriers. Climate change is a threat to human and ecological life, and the public is beginning to recognize it as such (Saad 2013). Thus, there is fertile ground for threat barriers to develop. In addition, potential solutions are often contested in political and social spheres, invoking goal barriers. The problem is also ambiguous, providing ample opportunity for these barriers to arise, as there is uncertainty or ambiguity concerning specific consequences of changing temperatures. For example, the Intergovernmental Panel on Climate Change (IPCC) Fifth Assessment Report stated that "there does not exist at present a single agreed on and robust formal methodology to deliver uncertainty quantification estimates of future changes in all climate variables" (IPCC 2007:1040). According to Chambers and Melkonyan (2017:74), this uncertainty has led to markedly different policy reactions, with some countries preferring to immediately tackle climate change, while others prefer to continue doing "business as usual." This ambiguity also impacts reactions at the personal level, as citizens are unclear what actions they should take (if any) to help mitigate impacts, thus also reinforcing goal barriers.

Lack of action at the personal level is also linked to ethical challenges that are unique to the problem of climate change. For example, Salomon et al. (2017) argues that one reason why people do little to modify their conservation behavior is that there is a disconnect between belief in climate change and behavioral responses. This lack of response can partially be explained by the popular belief that individual actions are nebulous and will do little to address the problem. It follows that they are not morally relevant. They call this phenomenon "climate change helplessness," or the belief that personal actions cannot impact climate change. Studies have shown that this feeling of helplessness in the face of enormous problems undermines peoples' motivations to change behaviors and, in this way, could short-circuit individually focused strategies.

In addition to motivational barriers, climate change helplessness highlights how this wicked problem brings about a "perfect moral storm," as it manifests three significant ethical challenges to action in a mutually reinforcing way (Gardiner and Hartzell-Nichols 2012). The first challenge concerns the scope of the problem, or the fact that climate change is a global phenomenon. Greenhouse gas emissions, once emitted, can cause effects anywhere on the planet (Shukla et al. 2018). This brings up a common environmental ethics problem, called the tragedy of the commons, although, in this instance, actors are nation states. While all countries should desire to limit global emissions, to reduce the risk of catastrophic impacts, it is in an individual's self-interest to continue emitting (e.g., Gardiner 2011; Helm 2008). This leads to a tragedy of the commons, where short-term gain undermines what is best for the collective. There is also the issue of skewed vulnerabilities, as the most vulnerable countries are those that have contributed the least to climate change, as they have low emissions. This creates a profound justice issue.

The second ethical challenge concerns how climate change has intergenerational effects. Greenhouse gas emissions persist in the atmosphere for long periods of time, contributing to negative impacts for centuries (IPCC 2007). This brings up ethical issues concerning fairness, and compounds barriers to cooperation, as mitigation strategies need to be coordinated across generations, and this seems unlikely. In addition, it is difficult to motivate nation-states and individuals alike to adopt austere changes for the sake of future generations.

The third and final ethical challenge is that we lack sophisticated social tools in relevant areas, such as intergenerational ethics, international justice, and environmental ethics (e.g., Jamieson 1992). For example, Gardiner and Hartzell-Nichols (2012) argue the following:

[C]limate change raises questions about the (moral) value of nonhuman nature, such as whether we have obligations to protect nonhuman animals, unique places, or nature as a whole, and what form such obligations take if we do . . . In addition, the presence of scientific uncertainty and the potential for catastrophic outcomes put internal pressure on the standard economic approach to environmental problems . . . and play a role in arguments for a precautionary approach in environmental law and policy that some see as an alternative.

There are several factors working together to hinder climate change mitigation strategies, including the three ethical challenges just discussed as well as motivational barriers and climate change helplessness. We've made great strides understanding barriers, the interdependencies between them, and the dynamic ways that they persist (Eisenack et al. 2014). Working to mitigate barriers is an important next step to ensure mitigation efforts are successful.

Climate change is a social problem

Today there is a growing literature on overcoming motivational barriers as well as ethical challenges. It is increasingly accepted that this research is imperative for developing effective strategies to adapt to fluxing temperatures. For example, Grundmann (2016) drew a comparison between the policy response to large-scale ozone loss over Antarctica, which occurred 30 years ago, and the contemporary lackluster response to climate change. He argues that a major reason for our failure to make substantive changes is that climate change is a social problem, yet it's framed as a scientific consensus issue. As such, the social realities have been downplayed or ignored in policy discussions. Previous research efforts stressed the need to better understand the phenomenon, as this is imperative for initiating policy and regulatory responses. This makes sense, as we should determine if there is a problem and, if so, the scope of the problem, before potential solutions are drafted. Thus, the organizations and experts involved in framing climate change as a problem and prescribing policy recommendations were predominantly trained in the natural sciences (Rayner and Caine 2014). Thus, they lack an awareness of human behavior and the process of social change. Indeed, the guidance of social scientists has "not been taken on board" (Grundmann 2016:562). If social scientists had been involved in a significant way from the beginning, the crucial error of categorizing climate change as a scientific consensus problem, rather than a social problem, could have been avoided.

If we reframe climate change as both a social problem and a wicked problem, then our goals shift. Rather than trying to "solve" the problem, we instead conceptualize it as a cluster of problems that need to "be re-solved and renegotiated, over and over again" (Grundmann 2016:562). Other social problems, such as reducing crime and maintaining healthy and educated populations, are all examples of concerns that do not offer one overall solution but are complicated and contextual. Success criteria are tied to political realities and subject to change. Indeed, when thinking about wicked problems, it's not realistic to imagine 100% success (a zero-crime rate, for example) as a feasible policy goal. The public knows this and so do policymakers. Goals are achieved incrementally, and social problems are managed better or worse, over time. In addition, cultural values play a key role, and what works in one place might not work in another. Thus, now that we have scientific consensus concerning climate change, climate science provides little help to meet this challenge.

Strategies to address motivational barriers

Today, there are several strategies to lessen the impact of the wicked nature of climate change, as well as motivational barriers. Duckett et al. (2016) provide a useful review of prominent strategies to tackle wicked environmental problems. As discussed, several of these are being employed in mitigation strategies, with promising results. First, the scholars identified strategies to help address the indefinability of wicked problems. This aspect of climate change gives rise to the ambiguity motivational barrier, as the nebulous nature of wicked problems hampers decisive action (Noll 2023). For instance, Berkes (2011) suggests that we should frame problems using a socioecological systems paradigm in order to prioritize these necessary components of wicked problems. Other scholars go further, arguing that we need to draw from environmental ethics to reimage technological and engineering practices, especially in relation to sustainability projects (Coyne 2005; Whyte and Thompson 2011). The latter is recommended to help tackle ethical challenges discussed earlier. These examples can be placed under the umbrella of proposals stressing the need for interdisciplinary or transdisciplinary approaches (Conklin 2010; Palmer 2012).

It is important to note that calls for interdisciplinarity stress the need for research by psychologists and communications experts on discourse, as discourses of fear could be harmful to mitigation efforts. In particular, Saab (2023:113) argues that "fear can lead to disengagement, 'climate change fatigue' and active opposition to climate change policies." It is imperative that we recognize that discourse of fears reflects not only sound climate science but also emotional rhetoric. Thus, the language that we use to discuss climate change is itself a challenge. Researchers, lawmakers and policymakers, and other individuals working to address climate change impacts "must acknowledge the rhetorical and emotive power of the language" and engage more seriously with research on discourse (p. 113). This is necessary for effective action, supported by stakeholders. Saab ends by recommending that we frame climate change using a comic apocalyptic narrative, rather than a tragic one. Comic narratives emphasize "provisionality of knowledge, free will, ongoing struggle and a plurality of social groups with differing responsibilities" (p. 132). This shift helps to mitigate the threat barrier, as it shifts our attention away from the tragic nature of climate change to, instead, embrace personal responsibility and the need for collective action. This is one example of ways that we can move away from discourses of fear. There are others in the literature, but it highlights how important socially focused research is for climate change strategies.

In addition to interdisciplinarity, other pragmatic strategies aimed at reducing ambiguity have been suggested. Duckett et al. (2016) identify two subthemes in this literature. First, there are several researchers who argue that we should adopt specific methodological approaches to help cope with uncertainty, such as scenario planning, modeling tools, and participatory approaches (see Batie 2008). Second, there are others who favor deconstructing wicked problems into subproblems that are easier to conceptualize and address (Shindler and Cramer 1999). A similar approach involves locking down definitions (Conklin 2010), maintaining focus on specific goals (Lazarus 2009), and finally, assessing competences and challenges at the local level. These more pragmatic approaches provide recommendations that help to mitigate the goal motivational barrier as well. The ambiguity of wicked problems could undermine adaptation and mitigation strategies, as they make it difficult to clearly define specific goals and endpoints. As discussed, when there are too many goals, they are contested, or they are ambiguous, individuals often fail to act quickly. Deconstructing wicked problems into subproblems, locking down definitions, and maintaining focus on specific tasks could address this barrier, as it replaces ambiguity with clearly defined action plans.

Overcoming barriers: steps forward

These and other strategies are being implemented with success, especially at the regional policy level. Governance efforts globally are developing slowly, as solving the climate problem is more difficult than anticipated (Bernauer 2013). This bottleneck has led to greater attention to bottom-up dynamics and support for regional and municipality responses (Bell et al. 2021). For example, state and local agencies are at the forefront of planning and responding to climate hazards, as weather patterns shift. Climate change, as a wicked problem, is complex, with success criteria that are subject to change and tied to the political realities of our time. As such, it makes sense that adaptation strategies would take the form of regional responses. The US Centers for Disease Control and Prevention and Climate-Ready States and Cities Initiative both provide funding to state and local health departments to respond to climate health impacts (Mallen et al. 2022). These public health initiatives use the Building Resilience Against Climate Effects framework, which has resulted in successful adaptation projects in Arizona, California, Oregon, and cities, such as San Francisco. Similarly, the USDA has developed a plan to mitigate the effects of climate change on agriculture, forests, and rural communities around the country. One of their goals is to reduce the vulnerability of human and natural systems and exposure to effects. They hope to accomplish this by creating Climate Hubs, or regional offices, to develop locally specific tools and resources to empower farmers, ranchers, and land managers to manage harms on the ground (USDA 2023).

The latter strategy is intended to help farmers mitigate impacts on the ground, as they experience them. Providing timely support could mean the difference between a successful and failed crop. This program could be very beneficial to ensure food security and for farming families around the country. Sustainable agriculture organizations, such as the Rodale Institute and Sustainable Agriculture Research and Education also provide support at the local level, as they help producers adopt environmentally friendly agricultural practices helpful for reducing climate change impacts. These include practices such as conservation tillage, the use of cover crops, diversified crop rotations, no-till methods, rotational grazing, and others. They can be effective methods for pulling carbon dioxide from the atmosphere and storing it in the soil.

Each of these examples of regional programs has identified a specific climate concern to focus on, has a clearly defined scope, and is working with local residents to address specific challenges faced by communities. Focusing on regional adaptation helps to clarify goals and bypass much of the ambiguity that tackling wicked problems brings, while also contributing to larger initiatives, such as reducing greenhouse gas emissions, etc. In addition to progress at the regional level, we are also seeing progress at the level of individuals. For example, as discussed, popular articles often claim that reducing your carbon footprint is a sound strategy to help contribute to mitigation efforts. However, even though people understand that climate change is a threat, individuals often do little to modify personal behavior when they feel that their actions will not make a difference (Solomon, Preston, and Tannenbaum 2017). While citizens have been slow to adopt energy conservation behavior, there is one area where individuals are embracing ecologically minded behavioral changes: green investing. This is an excellent example of a strategy that bypasses several of the motivational barriers discussed previously.

Green investing is a practice built on the idea that we can leverage investing to send market signals that carbon-polluting industries will not be tolerated. According to Voica et al. (2015), the main obstacle to achieving the goal of climate change mitigation is cost. There are several strategies that can be used to prompt industries to make this change, such as carbon taxes, but the authors go on to argue that an effective strategy is "the promotion of environmentally friendly investments, also known as green investments" (p. 72). According to Morgan Stanley's Institute for Sustainable Investing (2021), 79% of all investors reported interest in green investing. In 2016, assets worth $22.89 trillion (26% of all professionally managed assets) were managed using socially responsible and green investing strategies. Martini (2021) argues that there are two reasons for this growth, including the need for better corporate risk management practices and increasing public acceptance of a duty to change business practices to be more environmentally friendly.

This strategy is interesting, as it bypasses each of the barriers mentioned previously and leverages moral consideration to motivate behavioral change. First the message of green investing is tailored to reduce the feeling that we are threatened, as the activity is framed as a way to bring about larger economic change – change necessary to mitigate climate impacts. The root cause of climate change is also framed as an economic issue. This is reductionist and problematic, but it does reduce ambiguity. Second, green investing supporters distilled myriad possible solutions down into a single message. If you invest in green companies or those that prioritize environmental stewardship, conservation, and social responsibility, then you can generate long-term environmental benefits. While simplistic, the message provides individuals with a clear goal or way to discharge their perceived ethical duties, such as reducing carbon emissions, protecting the environment, improving the efficient use of resources, etc. While a prolonged evaluation and assessment of goals is an important part of adequately addressing wicked problems, providing an easily identifiable action point appears to be an effective strategy to marshal individualist responses. These could include the following: the recognition of a moral obligation to reduce individual emissions, the desire to lessen personal support for carbon-intensive industries by "voting with your dollar," a commitment to not participate in environmentally harmful group activities, etc. (Martini 2021). Each are individual actions, but they could help to bolster support for collectivist strategies, as citizens committed to making personal changes are more likely to support the adoption of regional and national environmental policies.

Society is struggling to respond effectively to our changing climate. This chapter explored how this wicked problem brings about unique social challenges – challenges that can act as motivational barriers hampering effective mitigation and adaptation. These include, but are not limited to, goal barriers, ambiguity barriers, and threat barriers, as well as ethical challenges curtailing immediate action. The social sciences and humanities provide insights into the social aspects of wicked problems, as well as strategies to address each of these barriers. This research is important if we hope to effectively motivate citizens and society to take steps to mitigate climate impacts. The aim of this chapter was to introduce readers to this important area of research. Understanding social barriers to action is imperative for effectively addressing what has been called the most wicked problem of the modern age: climate change.

References

Basseches, Joshua A., Rebecca Bromley-Trujillo, Maxwell T. Boykoff, Trevor Culhane, Galen Hall, Noel Healy, David J. Hess et al. 2022. "Climate Policy Conflict in the US States: A Critical Review and Way Forward." *Climatic Change* 170: 32.

Batie, Sandra S. 2008. "Fellows Address: Wicked Problems and Applied Economics." *American Journal of Agricultural Economics* 90: 1176–1191.

Baumeister, R. F. and Bargh, J. A. 2014. "Conscious and Unconscious: Toward an Integrative Understanding of Human Mental Life and Action." In *Dual-Process Theories of the Social Mind*, edited by J. Sherman, B. Gawronski and Y. Trope, 35–49. New York: Guilford Press.

Bell, James, Jacob Poushter, Moira Fagan and Christine Huang. 2021. "In Response to Climate Change, Citizens in Advanced Economies Are Willing to Alter How They Live and Work." *The Pew Research Center*. Accessed December 13, 2023. www.pewresearch.org/global/2021/09/14/in-response-to-climate-change-citizens-in-advanced-economies-are-willing-to-alter-how-they-live-and-work/

Berkes, Fikret. 2011. "Implementing Ecosystem-Based Management: Evolution or Revolution?" *Fish and Fisheries* 13: 465–476.

Bernauer, Thomas. 2013. Climate Change Politics. *Annual Review of Political Science* 16: 421–448.

Chambers, Robert G. and Tigran Melkonyan. 2017. "Ambiguity, Reasoned Determination, and Climate-Change Policy." *Journal of Environmental Economics and Management* 81: 74–92.

Cohen, Sheldon, Denise Janicki-Deverts and Gregory E. Miller. 2007. Psychological Stress and Disease. *JAMA* 298(14): 1685. https://doi.org/10.1001/jama.298.14.1685

Conklin, J. (2010). *Wicked Problems & Social Complexity*. CogNexus Institute Publications.

Coyne, Richard. 2005. "Wicked Problems Revisited." *Design Studies* 26(1): 5–17.

Duckett, Dominic, Diana Feliciano, Julia Martin-Ortega and Jose Munoz-Rojas. 2016. "Tackling Wicked Environmental Problems: The Discourse and Its Influence on Praxis in Scotland." *Urban Planning* 154: 44–56.

Eisenack, Klaus, Susanne C. Moser, Esther Hoffmann, Richard J. T. Klein, Christoph Oberlack, Anna Pechan, Maja Rotter and Catrien J. A. M. Termeer. 2014. "Explaining and Overcoming Barriers to Climate Change Adaptation." *Nature Climate Change* 4: 867–872.

Gardiner, Stephen M. 2011. *A Perfect Storm: The Ethical Challenge of Climate Change*. Oxford: University of Oxford Press.

Gardiner, Stephen M. and Hartzell-Nichols, L. 2012. Ethics and Global Climate Change. *Nature Education Knowledge* 3(10): 5.

Glazebrook, Tricia, Samantha Noll and Emmanuela Opoku. 2020. "Gender Matters: Climate Change, Gender Bias, and Women's Farming in the Global South and North." *Agriculture* 10(267): 1–25.

Grundmann, Reiner. 2016. "Climate Change as a Wicked Social Problem." *Nature Geoscience* 9(8): 562–563.

Heeter, Karen J., Grant L. Harley, John T. Abatzoglou, Kevin J. Anchukaitis, Edward R. Cook, Bethany L. Coulthard, Laura A. Dye and Inga K. Homfeld. 2023. "Unprecedented 21st Century Heat Across the Pacific Northwest of North America." *NPJ Climate and Atmospheric Science* 6(5).

Helm, Dieter. 2008. "Climate-Change Policy: Why Has So Little Been Achieved?" *Oxford Review of Economic Policy* 24: 211–238.

Intergovernmental Panel on Climate Change. 2007. *Climate Change 2007: The Physical Science Basis*. Cambridge: Cambridge University Press.

Jamieson, Dale. 1992. "Ethics, Public Policy, and Global Warming." *Science, Technology, and Human Values* 17: 139–153.

Jost, John T., Jack Glaser, Frank J. Sulloway and Arie W. Kruglanski. 2003. "Political Conservatism as Motivated Social Cognition." *Psychological Bulletin* 129(3): 339–375.

Kruglanski, Arie W. 2004. *The Psychology of Closed Mindedness*. London: Psychology Press.

Kruglanski, Arie W., Edward Orehek, E. Tory Higgins, António Pierro and Idit Shalev. 2010. "Modes of Self-Regulation: Assessment and Locomotion as Independent Determinants in Goal-Pursuit." In *Handbook of personality and self-regulation*, edited by R. Hoyle, 375–402. New York: Blackwell Publishing.

Lazarus, Richard J. 2009. "Super Wicked Problems and Climate Change: Restraining the Present to Liberate the Future." *Cornell Law Review* 94(5): 1153–1234.

Levin, Kelly, Benjamin Cashore, Steven Bernstein and Graeme Auld. 2012. "Overcoming the Tragedy of Super Wicked Problems: Constraining Our Future Selves to Ameliorate Global Climate Change." *Policy Sciences* 45: 123–152.

Mallen, Evan, Heather A. Joseph, Megan McLaughlin, Dorette Quintana English, Carmen Olmedo, Matt Roach, Carmen Tirdea, Jason Vargo, Matt Wolff and Emily York. 2022. "Overcoming Barriers to Successful Climate and Health Adaptation Practice: Notes from the Field." *International Journal of Environmental Research and Public Health* 19(12): 7169.

Martini, Alice. 2021. "Socially Responsible Investing: From the Ethical Origins to the Sustainable Development Framework of the European Union." *Environment, Development, and Sustainability* 23(1): 16874–16890.

Morgan Stanley Institute for Sustainable Investing. 2021. "Sustainable Signals: Individual Investors and COVID-19 Pandemic." https://www.morganstanley.com/assets/pdfs/2021Sustainable_Signals_Individual_Investor.pdf

Moser, Susanne C. and Julia A. Ekstrom. 2010. "A Framework to Diagnose Barriers to Climate Change Adaptation." *Proceedings of the National Academy of Sciences* 107(51): 22026–22031.

Noll, Samantha. 2017. "Climate Induced Migration: A Pragmatic Strategy for Wildlife Conservation on Farmland." *Pragmatism Today* 8(2): 17.

Noll, Samantha. 2018a. "Balancing Food Security and Ecological Resilience in the Age of the Anthropocene." In *Food, Environment, and Climate Change: Justice at the Intersections*, edited by E. Gilson and S. Kenehan, 179–193. Washington, DC: Rowman & Littlefield.

Noll, Samantha. 2018b. "Nonhuman Climate Refugees: The Role that Urban Communities Should Play in Ensuring Ecological Resilience." *Environmental Ethics* 40(2): 119–134.

Noll, Samantha. 2023. *From Food to Climate Justice: How Motivational Barriers Impact Distributive Justice Strategies for Change*, edited by F. Corvino and T. Andina, 271–289. Bristol: E-International Relations.

Palmer, James. 2012. "Risk Governance in an Age of Wicked Problems: Lessons from the European Approach to Indirect Land-Use Change." *Journal of Risk Research* 15(5): 495–513.

Pester, Patrick. 2021. *When Did Scientists First Warn Humanity About Climate Change?* Live Science. www.livescience.com/humans-first-warned-about-climate-change

Rayner, Steve and Mark Caine. (Eds.). 2014. *The Hartwell Approach to Climate Policy*. New York: Routledge.

Saab, Anne. 2023. Discourses of Fear on Climate Change in International Human Rights Law. *The European Journal of International Law* 34(1): 113–135.

Saad, Lydia. 2013. "Americans' Concerns About Global Warming on the Rise." *Gallup.Com.* https://news.gallup.com/poll/161645/americans-concerns-global-warming-rise.aspx

Salomon, Erika, Jesse L. Preston and Melanie B. Tannenbaum. 2017. "Climate Change Helplessness and the (De)moralization of Individual Energy Behavior." *Journal of Experimental Psychology* 23(1): 1–14.

Shah, James Y., Ron Friedman and Arie W. Kruglanski. 2002. "Forgetting All Else: On the Antecedents and Consequences of Goal Shielding." *Journal of Personality and Social Psychology* 83(6): 1261–1280.

Shalev, Idit. 2015. "The Climate Change Problem: Promoting Motivation for Change When the Map Is Not the Territory." *Frontiers in Psychology* 6: 131.

Shalev, Idit and Michael L. Sulkowski. 2009. "Relations Between Distinct Aspects of Self-Regulation to Symptoms of Impulsivity and Compulsivity." *Personality and Individual Differences* 47(2): 84–88.

Shindler, Bruce and Lori A. Cramer. 1999. "Shifting Public Values for Forest Management: Making Sense of Wicked Problems." *Western Journal of Applied Forestry* 14(1): 28–34.

Shukla, Priyadarshi R. et al. 2018. *IPCC, 2019: Climate Change and Land: An IPCC Special Report on Climate Change, Desertification, Land Degradation, Sustainable Land Management, Food Security, and Greenhouse Gas Fluxes in Terrestrial Ecosystems* [Governmental Website]. IPCC Climate Report. www.ipcc.ch/

Steele, Claude M. 1988. "The Psychology of Self-affirmation: Sustaining the Integrity of the Self." In *Advances in Experimental Social Psychology, Vol. 21: Social Psychological Studies of the Self: Perspectives and Programs*, 261–302. Cambridge: Academic Publishing.

Tollefson, Jeff. 2021. "IPCC Climate Report: Earth Is Warmer Than It's Been in 125,000 Years." *Nature* 596(7871): 171–172.

United Nations. 2021. "Actions for a Healthy Planet." https://www.un.org/en/actnow/ten-actions. Accessed March 07, 2024.

US Department of Agriculture. 2013. *Climate Change and U.S. Agriculture: An Assessment of Effects and Adaptation Responses.* Washington, DC: US Department of Agriculture.

US Department of Agriculture. 2023. *Climate Change Adaptation.* Washington, DC: US Department of Agriculture. www.usda.gov/oce/energy-and-environment/climate/adaptation

Voica, Marian Catalin, Mirela Panait and Irina Radulescu. 2015. Green Investments – Between Necessity, Fiscal Constraints and Profits." *Procedia Econoics and Finance* 22: 72–79.

Whitmarsh, Lorraine and Saffron O'Neill. 2010. "Green Identity, Green Living? The Role of Pro-environmental Self-identity in Determining Consistency Across Diverse Pro-environmental Behaviours." *Journal of Environmental Psychology* 30(3): 305–314.

Whyte, Kyle Powys and Paul B. Thompson. 2011. "Ideas for How to Take Wicked Problems Seriously." *Journal of Agricultural and Environmental Ethics* 25(4): 441–445.

8

WE ARE THE COLLECTIVE

Kristin Haltinner

The problem: Our societal stories about resolving the climate crisis focus on individual actions that, alone, will not make meaningful change in global emissions. Further, this narrative disincentivizes people from engaging in collective action that demands systemic and structural change. These stories also contribute to the paralyzing feelings of overwhelm experienced by concerned people.

The solution: We need a recognition that our actions can be and are part of collective action and that social change happens through restructuring the stories of what is possible. Through engaging our personal efforts in collective movements, we can change our cultural stories that perpetuate the climate crisis.

Where this has worked: Global movements such as Fridays for Future, Meat Free Mondays, 350.org, and Flygskam have engaged individuals in efforts that have had significant impact in changing cultural stories about the climate crisis, mobilizing individual and structural change.

My climate anxiety spiked after the birth of my son. Already exhausted from caring for an infant, I lay awake panicked about the future my son would endure as a result of the impending global destruction caused by climate change.

Easily influenced by dystopian fiction, the future I picture is vividly drawn in Octavia Butler's *Parable of the Sower* (Butler 1993). The novel depicts a society set in 2024 in the western United States, plagued by drought and rising sea levels,

DOI: 10.4324/9781003437345-12

economic destruction, and competition over food and land, led by a president who seeks to "make America great again."

Now, as I write this essay, massive flooding in Pakistan has resulted in nearly 1,500 deaths and widespread destruction (Peshimam and Hassan 2022). Moms, dads, children – like my own – and grandparents have lost their lives. Others, their homes. Here, in the Pacific Northwest, the air outside my window is "hazardous" due to the presence of smoke from regional forest fires – burning hotter and longer than ever before (Center for Climate and Energy Solutions 2022). We face a world in which conflicts over food and land are inevitable as food systems are weakened and climate migrants are forced to flee their homes in search of habitable land (Mbow et al. 2019). The world Butler envisioned is becoming a reality.

Almost unfathomably and despite repeated and unprecedented weather disasters, loss of human life, mass extinction of thousands of species, and the prediction of more to come, the international community, but especially the United States (the largest per capita contributor to elevated carbon dioxide [CO_2] levels), lacks the political will to take action on climate change.

Let me be clear – this inaction is *not* because we do not know what to do to prevent the full effects of the climate crisis. Countless nonprofits, governments, and international coalitions are clear and unified on what policies need to be enacted. For example, one of the leaders in political initiatives regarding climate, the UN Environmental Programme, offers a road map outlining six types of policy change needed to prevent global temperatures from increasing more than 1.5°C. These suggestions include revolutionizing buildings and cities, increasing reliance on renewable energy, and making changes to transportation, agriculture (e.g., reducing food waste), and industry operations while concurrently halting deforestation and environmental destruction (UN Environmental Programme 2022).

We know *what* needs to be done. The challenge is, rather, *how* we generate the public and political will necessary to enact adequate and meaningful climate protective policies.

The obstacles are numerous. We are a perversely and widely apathetic public. This apathy, explained clearly in Kari Marie Norgaard's book, *Living in Denial*, may be a response to the incredible overwhelm of climate disaster. In the face of the extraordinary threat, people preserve their psychological well-being by choosing to metaphorically "stick their heads in the sand" (Haltinner and Sarathchandra 2018; Norgaard 2011). The contradictory messages we receive about solving the crisis may further this tendency as they leave us confused, disoriented, and overwhelmed. We are told by news stories to focus on individual actions ("walk, don't drive") while being concurrently told that individual action doesn't matter given the massive emissions produced by industry. What are we to do?

Complicating the overwhelm and apathy among the public, we are saddled with psychopathically greedy politicians who benefit from the status quo and destruction of the planet. As a result, we stare down our destruction, surrounded by both a dearth of public will and political will to enact adequate and meaningful climate protective policies.

In this chapter, I explain the problems with the popular climate solutions offered and suggest a vision for engaging in collective action in support of climate change to enhance public will (and, in turn, political pressure) for climate policy.

The failure of current messages offered to solve climate change

Reading the newspaper drives any rational person to panic about the climate crisis. A brief look at CNN this week results in the following messages:

"Dangerous Heat Waves to at Least Triple Across the World by 2100, Study Says."
    "China's Worst Heat Wave on Record is Crippling Power Supplies."
    "Pakistan Floods Hit 33 Million People in Worst Disaster in a Decade, Minister Says."
    "Record-Breaking Heat Wave in Europe Will Be the Norm by 2035, Analysis Shows."

Once saddled with this anxiety and existential dread, caring people who want to contribute meaningfully to solving the climate crisis are then met with messages emphasizing their individual responsibility to change their lifestyle. For example, Googling, "how to solve climate change" results in the following suggestions:

"Save energy at home," "walk, bike, or take public transport," "eat more vegetables"
(from UN.org)

"Speak up," "Power your home with renewable energy," "weatherize, weatherize, weatherize," "invest in energy-efficient appliances," "reduce water waste," "actually eat the food you buy – and compost what you can't," "buy better bulbs," "pull the plugs," "drive a fuel-efficient vehicle," "maintain your ride," "rethink planes, trains, and automobiles," "reduce, reuse, and recycle"
(National Resources Defense Council 2022)

We are offered individual-level solutions to a global, institutional, and structural problem. These "solutions" emerge from an ideology deeply embedded in the American psyche – neoliberalism.

Neoliberalism, as a set of social policies, emerged in the 1970s (Harvey 2005). Its central arguments support the principles of free market competition (Harvey 2005). People who adhere to this theory believe that social change happens through market demands and that societal progress emerges only alongside economic growth. Neoliberalism advocates for limited state intervention in the realm of economics (Harvey 2005; Friedman and Friedman 1980; Friedman 1962).

It is this framework from which individual-level solutions to climate change are offered. It is also the root of suggestions for "green businesses" or, worse, arguments against industry regulations. For example, in the summer of 2022, the US Supreme Court restricted the ability of the Environmental Protection Agency to regulate carbon emissions without specific instruction and support of Congress (Totenberg 2022).

Agency, under neoliberal theory, comes from one's consumer choice. Individuals, then, have the freedom to choose their own preferences and relationships – personal or economic – into which they enter (Esposito and Finley 2014). We can choose to purchase electric vehicles or LED light bulbs.

Neoliberalism remains the dominant framework for economic, political, and social organization in the United States (Harvey 2005). Its focus on individual responsibility and consumer choice remains central to American capitalist ideologies and limits the possibility for structural and cultural change. It shapes the way we are allowed to engage with the climate crisis and limits our ability to envision adequate and needed solutions.

While on the one hand we are bombarded with messages about individual choices and behaviors needed to save the planet, on the other we receive the message that what we do as individuals won't solve the problem. Another list of recent headlines reflects this message:

"Individual Action Can't Stop Climate Change"

(Aspen Institute 2022)

"Individual Choices Won't Be Enough to Save the Planet"

(Time 2019)

This message recognizes the ways that major industries (e.g., animal agriculture) and corporations (e.g., ExxonMobil) dominate carbon emissions in the United States and that individual actions pale in comparison to change within these industries. Such narratives can lead people to overwhelm and inaction in the face of insurmountable problems and a lack of agency.

Faced with these contradictory messages, what are we to do?

Social change and the potential for a better world

Despite the dominance of neoliberalism in the United States, philosophers, sociologists, and other social theorists continue to assess and examine the factors that contribute to significant social change and their potential to solve the climate crisis. One place I find hope and direction in the resolution of social and environmental problems is in the work of postmodern feminists.

Postmodernism emerged in the mid-20th century as a movement to critique and assess the entirety of Western value systems (Lützeler 2001). One leader in this evolving school of thought was Michel Foucault. Foucault studied the ways that ideas and institutions develop and interact. His work seeks to provide witness to the ways taken-for-granted truths of a society are not inevitable or natural but simply the product of a particular historical trajectory.

According to postmodernism, within a given society, there are concurrent, competing knowledge systems. Even a dominant knowledge system operates parallel to others. In the case of climate change, we hear powerful narratives that challenge that status quo from people such as Greta Thunberg, whose messages often shame politicians and older adults for their inaction, leaving the suffering and responsibility for climate change to young people. Her Fridays for Future school protests, initially just her and a sign, spiraled across the globe, culminating in mass global protests in November 2015. In total, over 2,000 events were held in 175 countries. While Thunberg's message has not taken over as the dominant story about climate change in American society, it has significantly shifted public discourse and understanding of the climate crisis around the world.

From Foucault (2003), then, we see that ideologies are not absolute. They have what philosopher Sara Ahmed calls "leaks and spills" (Ahmed 2021a). These leaks can be capitalized upon by people to challenge existing norms and values (and their function in institutions) and create new knowledge systems. This process, however, is often slow. Philosopher Susan Bordo, for example, argues that change is rarely instantly radical or revolutionary in its alteration, saying "Such transformations do not occur in one fell swoop; they emerge only gradually, through local and often minute shifts in power" (Bordo 2004). According to Ahmed, these "minute" changes often look like "complaints" or saying "no." Ahmed argues that complaining is a method of pushing against power structures. It first serves to make visible power systems as they operate but also warns those who may encounter such systems in the future, all while challenging these structures (Ahmed 2021b; 2021c).

Let me give you an example of how this works in relation to climate change. In 2011, Swarthmore College became the first institution to divest from fossil fuels (Raji 2014). This effort came about due to pressure put on the college by a small group of (individual) students and faculty members working together to change policies in places over which they hold influence. A year later, Bill McKibben launched his 350.org campaign that encourages and supports the fossil fuel divestment campaigns that led to over 220 institutions joining the cause. In 2013 the Fossil Free campaign joined the effort in the United Kingdom, leading to additional divestments in Europe. Perhaps most astonishing was the 2014 announcement that the Rockefeller Brothers Fund would also divest from fossil fuel industries (Howard 2015). Within 4 years, the fossil fuel divestment movement grew to become the fastest growing divestment movement in history. Within 10 years, nearly 1,500 institutions, collectively reflecting nearly $40 trillion in assets, had committed to divesting from fossil fuels (Stand.Earth 2021). These campaigns remove investment funds from fossil fuel industries and businesses, which has led to an increase in capital costs. Companies themselves have indicated these increased costs to the divestment movement (McKibben 2018). Beyond simply decreasing the viability of reliance on fossil fuels, this movement has also succeeded in socially stigmatizing fossil fuel companies (Ansar et al. 2015). This implication further shows the ideological influences of action on climate change to shift the taken-for-granted assumptions rooted in cultures.

According to this theoretical framework, knowledge systems change in response to the production of counternarratives via saying no and making complaints. These counternarratives can exploit the leaks within dysfunctional systems resulting in new cultural norms and institutional policies. Social change happens, then, through concurrent changes in both systems of knowledge and social structures.

Unlike neoliberalism, postmodern feminism shows us that our action is part of a collective. While individual actions to fight climate change matter, they matter insomuch that they are part of a broader collection of voices making similar complaints and demanding similar change. Ahmed argues that this occurs through the building of what she calls "complaint collectives," groups of people (connected or disconnected) expressing shared subaltern knowledge systems (Ahmed 2021a). Ahmed contends that individuals – acting in concert – bring about change through their shared voice. As each "complaint" is made, it is "picked up and amplified by others." It expands and gains traction and force.

It is by capitalizing on this traction and bolstering complaint collectives that institutional policies change. Let me share another example. While initially aimed at animal rights, musician Paul McCartney and children Stella and Mary, began a Meat Free Monday campaign in 2009. Though not the first such campaign, McCartney's celebrity allowed it to catch on spreading to Brazil and Australia within the year. Since 2004, numerous newspaper columnists, cooking shows, and authors have given space to recipes for meatless Mondays. Journalist Michael Pollan emphasized the significance of the movement, saying that if every American went "one meatless day a week . . . that would be the equivalent of taking 20 million mid-size sedans off the road" (Oprah Winfrey Show 2009). Schools throughout the United States (including in New York City, Baltimore, Oakland, and elsewhere) and even cities (such as Aspen, Colorado) have instituted meatless Mondays. As a complaint collective, individual participation in meatless Mondays has had a significant impact on animal rights and carbon emissions. Moreover, approximately 33% of people who participate in meatless Mondays become a vegetarian within 5 years (De Visser et al. 2021). This suggests that the campaign also has allowed a subaltern knowledge system (that eating meat need not be central to one's diet) to emerge and expand.

Institutional change requires sustained efforts at saying "no," to identify the current invisible power operations, to warn others, and to push against them. Through the compilation of minute attempts into "complaint collectives," counter–knowledge systems are produced, and the momentum needed to shift discourse accordingly is maintained. As these fissures or "spills" expand, the ability to implement better and equitable policies, changing the institutional manifestation of power/knowledge in new ways is gained.

This ideological shift emphasizes the roles of individuals in collective action and gives weight to the influence of culture and discourse on institutional norms and behaviors. This perspective is reflected in part by social stories emphasizing the importance of "peer pressure," or "talking about it," to lead to both cultural change and shifts in institutional policy and practice.

What does fighting climate change look like?

In part, collective action for change does involve individual agency – not as an end, but as the beginning. Individual action must be viewed as part of the collective work to infiltrate dominant norms and practices to create something better.

Some of this is as simple as leading by example. In an experimental study, researchers Gregg Sparkman and his team found that patrons at a café in the United States were twice as likely to order a vegetarian lunch after being told that 30% of Americans had started eating less meat because of the climate crisis (Sparkman et al. 2020). Gordon T. Kraft-Todd and colleagues (2018) similarly found that people who installed residential solar panels themselves were 62.8% more likely to recruit others to do the same. These findings suggest that subaltern ideas and behaviors become increasingly possible and embraced when modeled by others.

While sometimes this peer pressure takes the form of modeling, in other cases it looks like collective shame. In Sweden, celebrities have publicly announced a commitment to giving up flying in exchange for more efficient modes of transportation. In turn, 23% of people in Sweden reduced their air travel (Timperley 2019). As a result, the airline industry has identified *Flygskam*, the "no fly" movement, as a significant threat to its business interest and has led some airlines, such as Jet Blue, to invest millions of dollars in carbon offsets (Hook 2019).

Here is the key factor – individual action matters in that an individual is part of collective action. To make individual action matter, then, one must also be vocal about the efforts they are taking in support of the planet, to influence the actions of others and amplify the voices of the collective. We need to change our cultural stories that accept climate change and our current extractive practices and replace them with one of collective responsibility. As Katharine Hayhoe says in her 2018 TED talk, "the most important thing you can do to fight climate change is to talk about it" and, specifically, talk about what you're doing to fight it. We must align ourselves with the work of others to reform institutions and say "no" to the dysfunctional operation of the industries destroying the planet.

And it works! Multiple efforts operate concurrently. They build steam and reach a tipping point in which counternarratives become dominant, and public will to change policy and institutional practices brings forth a tide of change. Consider, for example, the Civil Rights Movement. Though history often credits Martin Luther King Jr. for its success, the movement was made up of a constellation of complaint collectives seeking change in their local spaces. The Montgomery Improvement Association sought inclusion in bus transportation. The Student Nonviolent

Coordinating Committee fought for inclusion in public services and equal voting rights throughout the Southeast. Together these groups of individuals changed public opinion and popular will regarding equity, resulting in significant social change.

In this chapter, I've highlighted a few similar movements for climate action: divestment campaigns, Fridays for Future movement, and collective shifts toward vegetarianism. But there are more – WE ACT for Environmental Justice, Earth Guardians, Queer Nature, Honor the Earth, the Indigenous Environmental Network, The One Mind Youth Movement, and the Greta Thunberg Foundation, to name a few. These are collective efforts, with significant impact, made up by the decisions and choices of individuals. They are a network or constellation of efforts to solve the climate crisis.

To prevent global destruction, we need to expedite this practice and mobilize more people to join our complaint collective, demanding a radical restricting of political and industrial practices. We must change the stories that perpetuate an extractive and domination-based relationship with the earth and replace them with stories that emphasize collective responsibility and a shared ethic of care.

References

Ahmed, Sara. 2021a. "Feminists at Work." *Feminist Killjoys*. Accessed February 20, 2021. https://feministkilljoys.com/2020/01/10/feminists-at-work/

Ahmed, Sara. 2021b. *Complaint!* Durham, NC: Duke University Press.

Ahmed, Sara. 2021c. "Why Complain?" *Feminist Killjoys*. Accessed February 20, 2021. http://feministkilljoys.com/2019/07/22/why-complain

Ansar, Atif, Ben Caldecott and James Tilbury. 2015. "Stranded Assets and the Fossil Fuel Divestment Campaign: What Does Divestment Mean for the Valuation of Fossil Fuel Assets?" March 11. University of Oxford, Accessed December 12, 2023. www.smith-school.ox.ac.uk/sites/default/files/2022-03/SAP-divestment-report-final.pdf

Bordo, Susan. 2004. *Unbearable Weight*. Berkley: University of California Press.

Butler, Octavia. 1993. *Parable of the Sower*. New York: Four Walls Eight Windows.

Center for Climate and Energy Solutions. 2022. "Wildfires and Climate Change." Accessed September 16, 2022. www.c2es.org/content/wildfires-and-climate-change/

De Visser, Richard, Barnard, Suzanne; Benham, Daniel; Morse, Rachel. 2021. "Beyond 'Meat Free Monday': A Mixed Method Study of Giving up Eating Meat." *Appetite* 166: 105463. https//doi.org/10.1016/j.appet.2021.105463.

Esposito, Luigi and Laura L. Finley. 2014. "Beyond Gun Control: Examining Neoliberalism, Pro-gun Politics and Gun Violence in the United States." *Theory in Action* 7(2).

Foucault, Michel. 2003. *The Essential Foucault: Selections from Essential Works of Foucault, 1954–1984*. New York: The New Press.

Friedman, Milton. 1962. *Capitalism and Freedom*. Chicago: The University of Chicago Press.

Friedman, Milton; and Friedman, Rose. 1980. *Free to Choose*. New York: Harcourt Brace Jovanovich.

Haltinner, Kristin and Dilshani Sarathchandra. 2018. "Climate Change Skepticism as a Psychological Coping Strategy." *Sociology Compass* 12(6): e12586.

Harvey, David. 2005. *A Brief History of Neo-Liberalism*. Oxford/New York: Oxford University Press.

Hook, Leslie. 2019. "Year in a Word: Flygskam." *Financial Times*. December 29. Accessed December 13, 2023. www.ft.com/content/5c635430-1dbc-11ea-97df-cc63de1d73f4

Howard, Emma. 2015. "The Rise and Fall of the Fossil Fuel Divestment Movement." *The Guardian*. May 19. Accessed December 13, 2023. www.theguardian.com/environment/2015/may/19/the-rise-and-rise-of-the-fossil-fuel-divestment-movement

Kraft-Todd, Gordon, Bryan Bollinger, Kenneth Gillingham, Stefan Lamp. 2018. "Credibility-Enhancing Displays Promote the Provision of Non-normative Public Goods." *Nature*. 563(7730). https://doi.org/10.1038/s41586-018-0647-4

Lützeler, Paul. 2001. "From Postmodernism to Postcolonialism: On the Interrelation of the Discourses." *TRANS: Internet Zeitschrift für Kultuwissenchaften* 11. Accessed March 29, 2024. https://www.inst.at/trans/11Nr/luetzeler11.htm#

Mbow, Cheikh, Cynthia Rosenzweig, Luis Barioni, Tim Benton, Mario Herrero, Murukesan Krishnapillai, Emma Liwenga, Prajel Pradhan, Marta Rivera-Ferre, Tek Sapkota, Francesco Tubiello, and Yinlong Xu. 2019. "Food Security." In *Climate Change and Land: An IPCC Special Report on Climate Change, Desertification, Land Degradation, Sustainable Land Management, Food Security, and Greenhouse Gas Fluxes in Terrestrial Ecosystems,* edited by P. R. Shukla, J. Skea, E. Calvo Buendia, V. Masson-Delmotte, H.-O. Pörtner, D. C. Roberts, P. Zhai, R. Slade, S. Connors, R. van Diemen, M. Ferrat, E. Haughey, S. Luz, S. Neogi, M. Pathak, J. Petzold, J. Portugal Pereira, P. Vyas, E. Huntley, K. Kissick, M. Belkacemi, and J. Malley. Bonn, Germany: UNFCC.

McKibben, Bill. 2018. "At Last, Divestment Is Hitting the Fossil Fuel Industry Where It Hurts." *The Guardian*. December 16. Accessed December 13, 2023. www.theguardian.com/commentisfree/2018/dec/16/divestment-fossil-fuel-industry-trillions-dollars-investments-carbon

Natural Resources Defense Council. 2022. "How You Can Stop Global Warming." Accessed December 13, 2023. www.nrdc.org/stories/how-you-can-stop-global-warming

Norgaard, Kari. 2011. *Living in Denial*. Cambridge, MA: MIT Press.

Oprah Winfrey Show. 2009. "Eating Green." April 22. Accessed September 21, 2022. https://web.archive.org/web/20090428035106/www.oprah.com/article/oprahshow/20090422-tows-pollan-omnivore/1

Peshimam, Gibran and Hassan, Syed. 2022. "Death Toll in Pakistani Floods Nears 1,500; Hundreds of Thousands Sleep in Open." *Reuters*. Accessed September 16, 2022. www.reuters.com/world/asia-pacific/pakistan-floods-death-toll-nears-1500-2022-09-15/

Raji, Michelle. 2014. "Timeline: Fossil Fuels Divestment." *The Crimson*. October 2.

Sparkman, Gregg, Elizabeth Weitz, Thomas Robinson, Neil Malhotra and Gregory Walton. 2020. "Developing a Scalable Dynamic Norm Menu-based Intervention to Reduce Meat Consumption." *Sustainability* 12(6): https//doi.org/10.3390/su12062453

Stand.Earth. 2021. *Stand.earth*. October 26. Accessed December 1, 2021. https://stand.earth/insights/1485-institutions-with-assets-over-39–2-trillion-have-committed-to-divest-from-fossil-fuels/

Timperley, Jocelyn. 2019. "Why 'Flight Shame' Is Making People Swap Planes for Trains." *British Broadcasting System*. September 9. Accessed December 13, 2023. www.bbc.com/future/article/20190909-why-flight-shame-is-making-people-swap-planes-for-trains

Totenberg, Nina. 2022. "Supreme Court Restricts the EPA's Authority to Mandate Carbon Emissions Reductions." *National Public Radio*. June 30. Accessed December 13, 2023. www.npr.org/2022/06/30/1103595898/supreme-court-epa-climate-change

UN Environmental Programme. 2022. "Six Sector Solution." Accessed September 16, 2022. www.unep.org/interactive/six-sector-solution-climate-change/

9

INSIGHTS FROM SOCIAL AND BEHAVIORAL SCIENCES TO MOTIVATE CLIMATE ACTION

Dilshani Sarathchandra

The problem: While biophysical sciences have provided us with ample evidence for what causes climate change and what needs to happen to mitigate its impacts, individuals and societies often lack the motivation to act, an area in which we seem to struggle the most even in the midst of an unprecedented climate calamity.

The solution: Social and behavioral sciences provide us with valuable insights on factors that motivate individual and collective action to curb anthropogenic causes and impacts of climate change. These range from explorations of psychological factors such as affect and emotion that drive our individual decisions to group dynamics and cultural forces that shape our behavior such as identities and trust.

Where this has worked: Social science research suggests that varying degrees of pro-environmental attitudes exist, even among those who remain doubtful of climate change. They simply do not connect their environmental concerns with climate change. By leveraging these attitudes, we can quickly mobilize toward urgent action while simultaneously working to generate broader consensus among the public.

According to her Twitter biography, Greta Thunberg was "born at 375ppm."[1] Here, the young Swedish climate and environmental activist is referring to the global average atmospheric carbon dioxide (CO_2) level in 2003, the calendar year in which she was born. Fast forward to 2021, the global average CO_2 level has risen to 414.72 parts per million (ppm). According to the National Oceanic and

DOI: 10.4324/9781003437345-13

Atmospheric Administration's Global Monitoring Laboratory, this sets yet another "new record high" (Lindsey 2023). With every passing year, it seems climate predictions are becoming more dire. Referring to the 2022 Intergovernmental Panel on Climate Change report, UN Secretary General António Guterres claimed that our world is on "fast track to climate disaster" (*The Guardian* 2022).

Recognizing the gravity of this crisis at such a young age and refusing to look away, young Greta's efforts to fight climate change began at her home, with both her and her parents adopting lifestyle choices to reduce their own carbon footprint. In 2018 Greta initiated the now well-known "school strike for climate" movement that quickly spread into student climate protests throughout the world, collectively known as "Fridays for Future." Over the years, various other climate movements have grown in their membership, including the Sunrise Movement here in the United States, and the Debt for Climate Movement led by activists in the Global South, each, in its own way, pushing for much-needed, fast, and seemingly radical climate policy and action.

Among scientific and activist groups there is wide recognition that to curb the worst effects of climate change, we require concerted global, institutional, and structural reforms as well as widespread cultural and behavioral change. Public will must be mustered so that we are individually and collectively motivated to make significant lifestyle changes in the face of climate change, to pay more attention to it, to talk about it, and to vote into office those who are serious about pushing for and implementing policies that would lead to decarbonizing the economy. As my colleague Kristin Haltinner elaborates in Chapter 8, "individual action matters in that an individual is part of collective action." We are fast running out of time.

While biophysical sciences have aptly demonstrated the impacts of a changing climate on planetary sustainability, social and behavioral sciences provide us with valuable insights on factors that may motivate individual and collective action to curb the worst of its anthropogenic causes and effects. These range from psychological factors such as affect and emotion that drive our individual decisions to group dynamics and cultural forces that shape our behavior. In this chapter, I provide a brief overview of these factors as they pertain to climate change beliefs. For each topic under consideration, I highlight key insights from social and behavioral sciences along with their implications for climate change communication, policy, and action.

Risk perception

While experts use technical risk assessments to determine the probability and consequence of risk, such as when climatologists employ mathematical models to capture the effects of climate change, the general public often relies on their intuitive risk judgments. Psychologists have termed such intuition "risk perception" (Slovic 1987). While subjective, risk perceptions are important not only because they reflect how underlying factors such as our knowledge, experience, values, attitudes, and emotions influence our judgments about the seriousness and acceptability of risk, but also because they play a key role in motivating individuals to take action, or "to avoid, mitigate, adapt to, or even ignore risks" (Wachinger et al. 2013:1050).

When it comes to risk perception of natural disasters, prior research suggests that our personal experience, both direct experience via exposure to things like floods and wildfires, or indirect experience, via learning about these hazards through education and media, matters in important ways (Wachinger et al. 2013). For example, exposure to extreme climate or environmental events, especially when coupled with severe personal consequences, tend to amplify our risk perception which in turn could push us toward action. Indeed, we have seen some of these effects in our own research: we found that personally experiencing negative environmental events such as air and water pollution leads to increasing environmental concern and pro-environmental policy support, even among self-declared "climate change skeptics" (i.e., those who are doubtful or in denial about human-caused climate change) (Haltinner and Sarathchandra 2022).

However, like most things in life, the association between disaster exposure, risk perception, and our behavior isn't always straightforward. In their work on the "risk perception paradox," risk communication researcher Gisela Wachinger and her colleagues (2013) found that even individuals with high risk perceptions sometimes fail to take protective action. Elevated risk perception leads to risk-mitigating behavior, but only when certain conditions are met. People act when the benefits of action are seen as outweighing the costs of taking such action, when they realize their own agency to act, when the responsibility to act is not transferred to some other entity, and when they have sufficient resources to affect the situation.

In the context of climate change, this may mean that as more of us are personally, directly or indirectly, affected by more frequent and intense environmental/climate disasters, more of us will likely be amenable to changing our behavior, or at least to talking about it. However, we do not have time to wait for this to happen "naturally," nor do we want more people exposed to disaster. Yet, climate communicators may capitalize on these "windows of opportunity" after disaster events to engage the public, including those not directly affected, to gain their trust, and to motivate action. If you feel as though the 2023 summer of extreme climate events has shifted the discourse around climate change, created a sense of urgency, and woken more of us from our deep slumber, that indeed would be a step in the right direction.

Emotion and affect

Research in psychology, neuroscience, and related fields suggest that emotions and affect play important roles in individual decision-making and behavior. Prominent psychologist Paul Slovic and his colleagues define affect as "the faint whisper of emotion," where a feeling of "goodness" or "badness" is associated with external stimuli, often rapidly and automatically. People rely on affect when making judgments and decisions such that our decisions are driven by not only how we *think*, but also by how we *feel* (Slovic et al. 2004).

Emotions matter in that they motivate people to act, sometimes in ways we do not even recognize. Taking note of this power, some scholars have advocated for leveraging the potential motivating effects of negative emotions such as fear, anger, guilt, or regret to drive climate action, while others have prioritized prosocial emotions such as pride, gratefulness, and hopefulness (Haltinner and Sarathchandra 2018). Yet, exactly how and to what extent these distinct emotions motivate climate action remains inconclusive. For example, some argue that fear tactics, such as a focus on human extinction or economic devastation, may be paralyzing rather than motivating, resulting in denial or passivity (Ring 2015). Elsewhere, Kristin and I have argued that fear leads to information aversion, a presumed "ostrich effect," where we stick our heads in the sand rather than confront the issues at hand (Haltinner and Sarathchandra 2018).

Consequently, some scholars have recommended that rather than seeking the right "emotional recipe" for each climate change message, we should consider affect and emotion as part and parcel of a "broader, authentic communication strategy" (Chapman et al. 2017). Today, we see evidence of climate communicators, including scientists, journalists, and fiction/nonfiction writers, employing these various emotion-based rhetorical strategies in their own work such as when the American journalist David Wallace-Wells begins his 2019 book *The Uninhabitable Earth* with "It is worse, much worse, than you think," or, when George Mason University Professor Edward Maibach tweets, "The six key 'truths' about global warming: It's real; It's us; Experts agree; It's bad; Others care; & There's hope."[2]

Our own work demonstrates a complicated array of emotions when it comes to self-declared climate change skeptics. While skeptics experience less of an emotional charge around climate change, they express anger toward perceived antagonistic groups (e.g., scientists, liberals, environmentalists), but, at the same time, they worry about certain aspects of the environment that they care about – such as pollution, habitat loss, and species extinction – which we referred to as "objects of care," drawing from the work of Wang et al. (2018). To communicate with skeptics and convince them to act, it may be more fruitful to focus on these common areas of concern, emphasizing our individual and collective responsibility toward and our ability to protect these vulnerable objects (Haltinner, Ladino, and Sarathchandra 2021).

Identity and group dynamics

Persuading people that anthropogenic climate change is real and/or that they need to take action to reduce its effects gets complicated when we encounter intergroup dynamics such as identity. "Social identity theory," a prominent framework related to group dynamics, posits identity as a socially constructed self-categorization people use to define themselves as members of a group. Think about when one's political affiliation, religious beliefs, gender, or race generates seemingly distinct groups for us (Tajfel and Turner 1986). Social identities dictate common norms, values, beliefs, and attitudes for group members which facilitates ties between members. When a social identity becomes salient, an important part of how we each identify ourselves, we tend to treat our "in-group" members more favorably while developing potential antagonism toward members of the "out-group." This is the well-known "us versus them" dynamic we observe where everyday things like simply supporting different sports teams could sometimes generate strong negative feelings toward fans of the opposing team.

Social science scholarship suggests that these identity and intergroup dynamics are applicable to climate change in the sense that climate change skeptics and believers make distinct "opinion-based" social identities (Bliuc et al. 2015). Opinion-based social identities usually result from highly polarizing political attitudes where groups with opposing attitudes vie for social and political influence. As a result, climate change skeptics usually harbor a group consciousness that bonds them to each other but is also marked with animosity toward the opposing group (i.e., climate change believers).

In prior work, we characterized climate change skepticism as a "stigmatized, opinion-based social identity." The skeptics we interviewed in our research felt a sense of marginalization due to their perceived exclusion from mainstream society. Their climate beliefs ranged from outright denial to uncertainty and doubt, which correlated with other important determinants of beliefs and behavior such as emotional engagement, pro-environmentalism, engagement with science and media, and conspiracy ideation (i.e., the belief that climate change is a "hoax") (Haltinner and Sarathchandra 2023). Taking into consideration these identity dynamics, we proposed then that climate change skepticism is best understood as a "continuum" (Haltinner and Sarathchandra 2021). While not directly employing concepts from identity theory, researchers at the Yale Program on Climate Change Communication also divides the US public into what they call the "Six Americas," representing six distinct groups of shared thought: alarmed, concerned, cautious, disengaged, doubtful, dismissive (Leiserowitz et al. 2021).

Identity and group dynamics carry important implications for communicating climate change and motivating climate action. For instance, skeptics are likely to be receptive to messages stemming from their social "in-groups," such as conservative political and religious leaders. Framing climate stories in a way that activates shared identities held by both skeptics and believers (e.g., animal lovers, loving grandparents, being good people) may also be effective at decreasing the tendency in some to dismiss scientific evidence and information. Regardless of whether one is a skeptic or a believer, people are also likely to gravitate toward stories about others who are seen as similar to themselves in values, concerns, race, age, etc.

Tapping into pro-environmental attitudes

One common misconception about those who express denial or doubt about climate change is that they are "anti-environmental." In our research, we find that this is decidedly not the case. The self-declared skeptics we interviewed and surveyed over several years (2017 to 2020) expressed numerous pro-environmental views and concern about environmental degradation, yet they simply did not see those as connected to climate change.

Consider these facts: 62% of skeptics we surveyed in the Pacific Northwest region are concerned about air pollution and water quality; 59% worry about habitat destruction and deforestation; 55% worry about animal species loss; 70% support the preservation of national parks, and 58% even support the expansion of the National Park System. In interviews, these self-declared skeptics told us, "It's not like conservatives want to breathe dirty air!" (David); "There's danger in losing certain species of animals, which would be a tragedy in my mind" (Savannah); "Get away from more of these energy generators such as coal that pollute the environment and go into more sustainable ones" (Bill) (Haltinner and Sarathchandra 2020; 2022). For most skeptics, we found that these pro-environmental attitudes are connected to their lived experiences such as vivid childhood memories in outdoor spaces or personal negative health impacts of toxic exposure.

Commonly shared pro-environmental attitudes among skeptics and believers have important implications for communication, policy, and action. Take air pollution, for one. Air pollution results mainly from the burning of fossil fuels (coal, oil, gas). The science is undisputed that air pollution leads to severe negative consequences for human health. Many air pollutants are also heat-trapping greenhouse gases that damage our climate's health (Mackenzie 2023).

This is an issue that we all care about, an area where there is promise for collective action through increasing pressure on our elected leaders to enact policy and strengthening existing policy (e.g., Clean Air Act; the US Environmental Protection Agency's (EPA) mandate to monitor and regulate the air quality). The same is true for other issues such as habitat loss, deforestation, species loss, and preservation of national parks – all issues that people care about across the board; all issues when addressed will increase pathways for carbon sequestration. It is essential that those who enact policy, especially Republican policymakers, are made aware of the wider public desire and demand in these areas such that their legislative and funding decisions align better with the priorities of their constituents (Haltinner and Sarathchandra 2023).

Recently, climate change scholars and communicators have been advocating for "rebranding" carbon emissions as "pollution," essentially defining climate change as an air pollution problem. Terms such as "carbon pollution" and "climate pollution" are growing in popularity, an approach that has received support from prominent climate communicators such as Susan Hassol, the director of Climate Communication, a science outreach nonprofit, who once stated to the *Grist* magazine: "I think 'pollution' is a better word to use than 'emissions,' because everyone understands that pollution is harmful" (Yoder 2022). Positioning climate change as a pollution problem has wide-ranging implications for both individual attitude and behavior change as well as for enacting policy. As opposed to "carbon emissions" which is abstract for most of us and which none of us can see, "carbon pollution" brings up visceral images, helping us more easily connect two issues that are inextricably linked in the real world – global carbon emissions and local air pollution. Institutionally, the recent amendment to the Clean Air Act which categorizes greenhouse gas emissions as a form of air pollution in Biden administration's 2022 Inflation Reduction Act allows the EPA to continue to regulate carbon emissions (Hijazi 2022).

Information systems and trust

Climate change is a politically charged topic in the United States. Its politicization has resulted in significant polarization of public attitudes toward how we should respond to it. For instance, Pew Research Center's data show that while 90% of Democrat/Democrat-leaning US adults think that the federal government is "doing too little to reduce the effects of climate change," only 39% of Republican/ Republican-leaning Americans agree with this sentiment (Funk and Hefferon 2019). In fact, decades of social science scholarship have reported on attitude polarization on climate change, revealing that men, individuals who are politically conservative, older adults, and those who are more religious tend to be more skeptical of climate change than their respective counterparts (McCright and Dunlap 2011).

Climate change attitude polarization has not occurred in a vacuum. Social scientific analysis reveals that a concerted, long running, and financially rich "climate change denial countermovement," consisting of conservative think tanks, members of the fossil fuel industry, Republican politicians, contrarian scientists, and conservative media, has been especially savvy at sowing doubt about climate change in the United States (Dunlap and McCright 2015). To do this, the countermovement has aligned its messages with certain elements of the conservative political ideology, positioning climate action as a threat to conservative political goals (e.g., small government, free enterprise) (Collomb 2014). It is not hard then to imagine how this framing may have activated an "us vs. them" identity dynamic as discussed earlier, putting skeptics in a defensive position against any meaningful climate action.

Not only have activists representing the denial countermovement successfully generated a sense of threat among skeptics, but their activism constitutes cultural forces that have effectively siloed us in our own information systems (i.e., echo chambers). Partisan media exacerbates this effect via one of our common cognitive biases known as "motivated reasoning" that leads to the rejection of new information and intensification of prior views when encountering new or contradictory information (Hart and Nisbet 2012). Scholars have identified political ideology and party identification as the primary causes of motivated reasoning (Zhou 2016). These effects are likely prominent among individuals whose skeptic identities have been activated via messaging emanating from the denial countermovement.

Mis/disinformation circulating in echo chambers is especially hard to combat as it deliberately targets identities, ideologies, heuristics, and biases of audience members. For decades now, denialists have used rhetorical strategies to muddy the water regarding climate science (Cook 2020). For instance, consider the belief that climate change is a "hoax." While scientists have reached unanimity in their consensus that climate change is occurring, this belief continues to percolate in social media including when former president Donald Trump, prior to ascending to presidency, once tweeted "Global warming is an expensive hoax!"[3] In our research, we found that those who adhere to this belief often blame the United Nations, global elites, the EPA, democratic political leaders, and climate scientists for perpetuating this myth. They are less trusting of mainstream scientific, political, and environmental organizations but hold positive views toward organizations and leaders representing the political right (e.g., Republican leaders, Fox News). Demographically, they tend to be predominantly male, older, and more politically and religiously conservative than those who do not believe that climate change is a hoax (Sarathchandra and Haltinner 2021). Skeptics are especially distrusting of climate scientists whose work (and the resulting scientific narrative) is seen as the main driver of climate policy that threatens their own beliefs and values. As a result, skeptics question scientists on several grounds including for presumed lack of adequate data, for being exclusionary, and/or for being driven by certain tenure/career pressures (Sarathchandra et al. 2022).

To overcome these powerful institutional and cultural forces working against climate action, we need to tap into many of the social, psychological, and behavioral processes discussed earlier. We need to leverage personal experiences, such as sharing firsthand accounts of climate/environmental disasters or mimicking direct experience via other media (e.g., fiction, documentaries, film, virtual reality, augmented reality). We need messaging about climate change and climate action to prioritize our shared identities and our emotional connections to human and nonhuman beings. We need to immediately push for policy changes where wide public demand for such change already exists. Communicating the scientific consensus around the issue and rebuilding trust between experts and the public also become important considerations here. In fact, recent research suggests that communicating the scientific consensus is one of the most promising interventions to close the expert-lay gap in public's belief in scientific facts regarding climate change (Stekelenburg et al. 2022). Climate scientists, journalists, and science writers would do well to highlight the scientific consensus around climate change rather than continue to cover contrarian views for the sake of "balance." Climate change is happening now, and we need to act in the here and now.

Notes

1 @GretaThunberg on X/Twitter.
2 @MaibachEd, Nov 6, 2022, on X/Twitter.
3 Former President Donald Trump tweeted this on January 19, 2014. The full tweet read, "Snowing in Texas and Louisiana, record setting freezing temperatures throughout the country and beyond. Global warming is an expensive hoax!"

References

Bliuc, Ana-Maria, Craig McGarty, Emma F. Thomas, Girish Lala, Mariette Berndsen and RoseAnne Misajon. 2015. "Public Division About Climate Change Rooted in Conflicting Socio-political Identities." *Nature Climate Change* 5(3): 226–229.

Chapman, Daniel A., Brian Lickel and Ezra M. Markowitz. 2017. "Reassessing Emotion in Climate Change Communication." *Nature Climate Change* 7(12): 850–852.

Collomb, Jean-Daniel. 2014. "The Ideology of Climate Change Denial in the United States." *European Journal of American Studies* 9: 9–1.

Cook, John. 2020. "Deconstructing Climate Science Denial." In *Research Handbook on Communicating Climate Change*, 62–78. Cheltenham: Edward Elgar Publishing.

Dunlap, Riley E. and Aaron M. McCright. 2015. "Challenging Climate Change." In *Climate Change and Society: Sociological Perspectives*, 300. Oxford: Oxford University Press.

Funk, Cary and Meg Hefferon. 2019. "U.S. Public Views on Climate and Energy." Retrieved November 15, 2022. www.pewresearch.org/science/2019/11/25/u-s-public-views-on-climate-and-energy/

The Guardian. 2022. "World on 'Fast Track to Climate Disaster', Says UN Secretary General." Retrieved November 15, 2022. www.theguardian.com/environment/video/2022/apr/04/world-on-fast-track-to-climate-disaster-say-un-secretary-general-video

Haltinner, Kristin and Dilshani Sarathchandra. 2018. "Climate Change Skepticism as a Psychological Coping Strategy." *Sociology Compass* 12(6): e12586.

Haltinner, Kristin and Dilshani Sarathchandra. 2020. "Pro-environmental Views of Climate Skeptics." *Contexts* 19(1): 36–41.

Haltinner, Kristin and Dilshani Sarathchandra. 2021. "Considering Attitudinal Uncertainty in the Climate Change Skepticism Continuum." *Global Environmental Change* 68: 102243.

Haltinner, Kristin and Dilshani Sarathchandra. 2022. "Predictors of Pro-environmental Beliefs, Behaviors, and Policy Support Among Climate Change Skeptics." *Social Currents* 9(2): 180–202.

Haltinner, Kristin and Dilshani Sarathchandra. 2023. *Inside the World of Climate Change Skeptics*. Washington, DC: University of Washington Press.

Haltinner, Kristin, Jennifer Ladino and Dilshani Sarathchandra. 2021. "Feeling Skeptical: Worry, Dread, and Support for Environmental Policy Among Climate Change Skeptics." *Emotion, Space and Society* 39: 100790.

Hart, P. Sol and Erik C. Nisbet. 2012. "Boomerang Effects in Science Communication: How Motivated Reasoning and Identity Cues Amplify Opinion Polarization About Climate Mitigation Policies." *Communication Research* 39(6): 701–723.

Hijazi, Jennifer. 2022. "Climate Law Gives Clean Air Act a Legal Boost After Court Rebuke." Retrieved November 15, 2022. https://news.bloomberglaw.com/environment-and-energy/climate-law-gives-clean-air-act-a-legal-boost-after-court-rebuke

Leiserowitz, Anthony, Connie Roser-Renouf, Jennifer Marlon and Edward Maibach. 2021. "Global Warming's Six Americas: A Review and Recommendations for Climate Change Communication." *Current Opinion in Behavioral Sciences* 42: 97–103. https//doi.org/10.1016/j.cobeha.2021.04.007

Lindsey, Rebecca. 2023. "Climate Change: Atmospheric Carbon Dioxide." Retrieved June 12, 2023. www.climate.gov/news-features/understanding-climate/climate-change-atmospheric-carbon-dioxide

Mackenzie, Jillian. 2023. "Air Pollution: Everything You Need to Know." National Resource Defense Council. Retrieved November 24, 2023. www.nrdc.org/stories/air-pollution-everything-you-need-know

McCright, Aaron M. and Riley E. Dunlap. 2011. "Cool Dudes: The Denial of Climate Change Among Conservative White Males in the United States." *Global Environmental Change* 21(4): 1163–1172.

Ring, Wendy. 2015. "Inspire Hope, Not Fear: Communicating Effectively About Climate Change and Health." *Annals of Global Health* 81(3): 410–415.

Sarathchandra, Dilshani and Kristin Haltinner. 2021. "How Believing Climate Change is a "Hoax" Shapes Climate Skepticism in the United States." *Environmental Sociology* 7(3): 225–238.

Sarathchandra, Dilshani, Kristin Haltinner and Matthew Grindal. 2022. "Climate Skeptics' Identity Construction and (Dis)trust in Science in the United States." *Environmental Sociology* 8(1): 25–40.

Slovic, Paul. 1987. "Perception of Risk." *Science* 236: 4799.

Slovic, Paul, Melissa L. Finucane, Ellen Peters and Donald G. MacGregor. 2004. "Risk as Analysis and Risk as Feelings: Some Thoughts About Affect, Reason, Risk, and Rationality." *Risk Analysis* 24: 311–322.

Stekelenburg, Aart van, Gabi Schaap, Harm Veling, Jonathan van't Riet and Moniek Buijzen. 2022. "Scientific-Consensus Communication About Contested Science: A Preregistered Meta-Analysis." *Psychological Sciences* 33(12): 1989–2008.

Tajfel, Henri and John Turner. 1986. "Social Identity Theory of Intergroup Behavior." In *Psychology of Intergroup Relations*, edited by Stephen Worchel and William Austin, 7–24. Chicago: Nelson-Hall.

Wachinger, Gisela, Ortwin Renn, Chloe Begg and Christian Kuhlicke. 2013. "The Risk Perception Paradox – Implications for Governance and Communication of Natural Hazards." *Risk Analysis* 33: 1049–1065.

Wallace-Wells, David. 2019. *The Uninhabitable Earth: Life After Warming*. New York: Tim Duggan Books, 1st edition.

Wang, Susie, Zoe Leviston, Mark Hurlstone, Carmen Lawrence and Iain Walker. 2018. "Emotions Predict Policy Support: Why It Matters How People Feel About Climate Change." *Global Environmental Change* 50: 25–40.

Yoder, Kate. 2022. "It Makes Climate Change Real." Retrieved November 15, 2022. https://grist.org/health/how-carbon-emissions-got-rebranded-climate-pollution-ira/

Zhou, Jack. 2016. "Boomerangs Versus Javelins: How Polarization Constrains Communication on Climate Change." *Environmental Politics* 25(5): 788–811.

10

ACT NOW

Words, actions, and values in tackling the human dimensions of the climate crisis

Jack DeWaard

The problem: The world is standing on the precipice of a climate catastrophe and, especially in more political and politicized circles, its human inhabitants are unwilling and/or unable to have honest conversations and chart corrective directions forward for the good of the planet and its peoples.

The solution: Individual and collective words, actions, and values are instrumental, to start and work from, as we approach, tackle, and ultimately solve the climate crisis.

Where this has worked: In their efforts to date to propose and manage a new marine sanctuary along California's central coast, the Chumash Tribe has led by example and grounded their work in a clearly articulated and forward-looking set of words, actions, and values.

The climate crisis has been called many things: global warming, climate change, climate disruption, and climate emergency, to name a few. The changing terms we use to discuss this crisis are, at times, an attempt by scientists to correctly label the urgent disasters we face. In other uses, these terms seek to downplay or ease concerns or tensions that result from the ways in which climate change has become polarized. On too many occasions, climate scientists have been prohibited from using accurate scientific descriptors of the crisis. Other times they self-censor in response to political moves, afraid that using correct terminology will result in backlash and preferring, instead, to use phrases such as "extreme weather" (Hersher 2017).

DOI: 10.4324/9781003437345-14

This censorship of discussion of climate change distorts the urgency of the crisis while concurrently preventing a robust and honest conversation about the emergency we face. This forced distortion results in an inability for citizens to engage in a movement toward action to prevent the worst outcomes of climate disruption.

In this context, this chapter highlights the importance of individual and collective words, actions, and values in approaching, tackling, and ultimately solving the climate crisis. While these may seem innocuous enough, they are anything but. In a world that is standing on the precipice of a climate catastrophe and with its human inhabitants unwilling and/or unable to have honest conversations or chart corrective directions forward for the good of the planet and its peoples (especially in some more political and politicized circles), collective engagement is essential. I close with a recommendation and challenge to "ACT now." Using the example of the efforts of the Chumash Tribe to propose and manage a new marine sanctuary along California's central coast, I recognize and celebrate the need to ACT. The acronym, ACT, is borrowed from Acceptance and Commitment Therapy, which emphasizes two key steps that include individually and collectively 1) acknowledging and accepting the scary realities and uncertainties associated with the worsening climate crisis and 2) committing to a values-driven course of action that is consistent with who we are and who we want to be in the face of these catastrophic changes.

Recognizing the problem

The other day, I was on a phone call catching up with a friend who lives in Washington State. After our usual back-and-forth exchanging updates about all things work and life, I said something to the effect of, "So, what else is new?" My friend responded by saying that he had recently driven down to California's San Joaquin Valley to help his sister pack up her house and move to Washington State. As someone who studies human migration, I am quite familiar with the constant trickle of Californians moving up to Washington State and the Pacific Northwest more broadly; however, when I followed up with a question about the reason(s) for my friend's sister's move, I heard something that I hadn't really heard before. My friend said that his sister who is disabled and lives alone on a fixed income could no longer keep up with rising costs of cooling her San Joaquin Valley home during the summer, a season that has seen temperatures consistently high and remain well above 100°F for days and even weeks on end. So, rather than risk losing her home, my friend's sister threw in the towel and made the trek north to Washington State.

About the same time, in November 2022, the Biden administration made headlines by announcing $75 million in new funding for three Native American tribes in Alaska and Washington State to support the voluntary relocations of their communities to higher ground away from coastlines and rivers that have risen and will continue to rise with climate and environmental change (US Department of the Interior 2022). In many ways, just like my friend's sister, members of these communities can and should be considered climate and "environmental refugees," a term that first appeared in a 1985 report by the UN Environment Programme and was defined as "people who have been forced to leave their traditional habitat, temporarily or permanently, because of a marked environmental disruption (natural and/or triggered by people) that jeopardized their existence and/or seriously affected the quality of their life" (El-Hinnawi 1985).

However – and something that really took me by surprise when I first encountered this – prominent organizations like the UN High Commissioner for Refugees (UNHCR) and the International Organization for Migration (IOM) actually recommend "to avoid terms such as climate change refugee and environmental refugee" (IOM 2014). Their rationale is that terms like *climate* and *environmental refugee* have the potential to undermine and erode the legal definition of a refugee who was established after World War II. The 1951 *United Nations Convention Relating to the Status of Refugees* defined a refugee as a person who, "owing to well-founded fear of being persecuted for reasons of race, religion, nationality, membership of a particular social group or political opinion," is unable or unwilling to return to their country of origin (UN General Assembly 1951). Two details are particularly important here. First, a refugee must be *persecuted by* some sort of entity, which is often envisioned as a government or government actor. Second, a refugee must be persecuted *on the basis of* one or more of the five grounds (e.g., race) listed earlier. Organizations like the UNHCR and the IOM argue that terms like *climate refugees* and *environmental refugees* should not be used because climate and environmental change is not an entity and because the impacts of climate and environmental change do not discriminate based on any of the five grounds listed earlier.

This is shortsighted for at least two reasons. First, with respect to being persecuted by some sort of entity, the evidence is clear that the world's largest carbon dioxide (CO_2)–emitting countries, a group that includes the United States, shoulder the bulk of responsibility for the climate crisis and are the wealthiest nations, a fact that was on full display and a point of contention and negotiation at the 2022 UN Climate Change Conference, or the 27th Conference of the Parties (COP27), in Egypt (Hickel 2020; Plumer et al. 2022). To make matters worse, none of the world's largest CO_2-emitting countries are on track to meet emissions targets that they set for themselves and signed on to keep in the Paris Agreement following COP21 in Paris (Bearak and Popovich 2022).

Second, with respect to being persecuted on the basis of one or more of the five grounds listed earlier, there is also a strong case to be made that this has happened and continues to happen during the climate crisis. To see this, consider the contrasting stories of Andre and Jim who were each homeowners in New Orleans in early 2005 (Gabour 2015). As you might recall, later that year, Hurricane Katrina made landfall and remains the costliest hurricane on record (National Centers for Environmental Information 2021). Given differences in the wealth levels and the locations of their homes, the designs, and the construction materials used, Andre's home was much more badly damaged by Hurricane Katrina than Jim's.

Andre's home was located on a floodplain, lacked critical design features such as stormproof windows, and was constructed with subpar materials. At the end of the day, then, the natural disaster that was Hurricane Katrina was anything but natural because it merely exposed preexisting differences between Andre and Jim, at least with respect to the features of their homes.

In thinking about climate and environmental refugees being persecuted on the basis of one or more of the five grounds listed earlier, the key is to realize that preexisting inequalities are often organized along lines that strongly overlap these five grounds. Hurricane Katrina, for example, laid bare significant preexisting inequalities by race and wealth in and around New Orleans (DeWaard 2020). Similarly, small island states like Kiribati face possible extinction from rising seas that they did not cause, thereby raising questions about persecution on the grounds of nationality by those like the United States and other major emitters who are primarily responsible for this problem (Millman 2022). The list goes on and on, but, at the end of the day, the key message here is that by hiding behind antiquated and narrow legal definitions of who qualifies as a refugee, we run the serious risk of tunnel vision in a world that is constantly and increasingly changing in ways that require more expansive and accurate views and acknowledgments of persecution and of the bases of persecution (Gemenne 2015).

Of course, the scope of my comments here is not limited to climate and environmental refugees. In many ways, questioning whether and how the 1951 definition of a refugee should be expanded to also include those impacted by climate and environmental change is just one manifestation of broader discussions and debates on how to think about, and what to do about, the human dimensions of climate and environmental change. The case of environmental migrants and the denial to recognize their status as refugees distorts the experiences of these peoples and hides the culpability of others. By failing to use accurate language in our discussion of the climate refugee crisis – just like climate change itself – we are unable to fully engage in an understanding of the root causes of the problem and, subsequently, move forward with informed solutions.

In what follows, I suggest that three fundamental things matter and deserve more focused attention going forward in our efforts to fight the climate crisis: our words, our actions, and our values.

Words

Our words matter.

Several years ago, I had the opportunity to speak about some of my research on climate and environmental migration and displacement at an unnamed university in an unnamed place in front of an audience that included an unnamed politician. Minutes before I was about to begin speaking, one of the event organizers approached me and in a hushed and worried tone asked, "You're not going to use the term, 'climate change,' are you?" Somewhat taken aback, I responded by telling the organizer that I would be using this term, not to be controversial or political, but because the science is abundantly clear that climate change is real and very likely due to human activity (National Aeronautics and Space Administration 2022).

When I recall this exchange, it reminds me of a conversation that I recently had with my daughter who has cerebral palsy. She and I were talking about the growing use of the term *differently abled* in place of the term *disabled*. Coming into this conversation, I think I expected my daughter to favor the former term given the lack of deficit-based framing and the emphasis on difference and diversity. However, she strongly advocated for the latter term. Her argument was simple: in the world in which she finds herself living every day, wherein ableism is unfortunately the norm, the term *disabled* is simply more accurate and honest. Using the term disabled more accurately points to the way that society denies access and support to people who are viewed as different and, thus, separates social responsibility for improving accessibility in society.

By this logic, I think we can and should say the same about climate change, which is not merely a change but a full-fledged crisis that is an existential threat to the planet and its peoples (Ripple et al. 2020). When we refuse to use words like crisis, emergency, or disruption, we play down the impending disaster. In our efforts to increase people's comforts, we, instead, remove important contexts from our discussions of climate change. In easing people's minds, we distance ourselves from the human dimension of climate change and the impending disasters that will unfold without action. This is why my colleagues and I at the Population Council use words like *crisis* in the title of one of our climate research initiatives – the Population, Environmental Risks, and the Climate Crises initiative (PERCC). Beyond the fact that words such as *crisis* accurately describe the gravity of the situation that we are currently facing, they also convey a sense of urgency, which raises subsequent questions about our corresponding actions.

These examples – the distorted language around climate refugees, denial of the climate crisis, and framing of disability, all serve as examples to demonstrate the hoops we jump through to find comfort in the face of that which is and should be uncomfortable. In each case, we perform linguistic acrobatics that serves to separate real social crises from the root causes and, thus, prevent our ability to find solutions.

Actions

Our actions matter.

While this might seem obvious, I mean this in a specific way. To illustrate, let me describe one of the main ways that I think about and approach my research on migration. When I was in graduate school and starting to take an interest in migration, I grew fascinated with how researchers trained in geography approached migration in very spatially explicit ways that prioritized how places, and people and populations living in them, are connected to one another by migration.

A couple years later, my colleagues and I applied this spatial thinking to study migration from and to New Orleans and other places in the US Gulf Coast before and after Hurricane Katrina (Curtis et al. 2015; DeWaard et al. 2016; Fussell et al. 2014). One of the things that became abundantly clear from this research was that many places in the United States that were not directly impacted by Hurricane Katrina were nonetheless indirectly impacted by this event through migration and, as such, had and continue to have a vested interest in what happens to places impacted by extreme weather events.

Applied to discussions and debates about the human dimensions of climate and environmental change, including climate and environmental refugees, I think it is critical to bring these sorts of interconnections and interdependencies between people, populations, and places to the surface for at least two reasons. First, doing so helps us to acknowledge and appreciate that our individual and collective actions (e.g., CO_2 emissions, as discussed earlier) affect one another in both direct and indirect ways over time and space. Second, by acknowledging and appreciating that our actions affect one another, this creates space to see our individual and collective fates as necessarily intertwined such that, contrary to some sort of zero-sum game wherein my loss is your gain, we stand to reap the largest gains when we work together for the collective good.

While this might look and sound nice, in reality, we often fail to appreciate that our actions matter in the ways that I described earlier. For instance, take Chinese leader Xi Jinping's comments in July 2023 about China going it alone on greenhouse gas emissions without foreign interference (Shepherd et al. 2023). These sorts of isolated actions are the antithesis of what I described earlier and what we need to prevent the worst impacts of the climate crisis. Rather, we need informed collective action – rooted in the knowledge that we, together, have caused this crisis and that only we, together, can solve it.

Isolated efforts to fight climate change also serve as a reminder of why we cannot be content to stop with our words and actions. Instead, we must go the next step to also interrogate and clarify our underlying and guiding values.

Values

Our values matter.

Of course, underlying any commitment to accurate and honest words and interconnected and interdependent actions is a set of values that need to be made explicit. For example, in the story that I shared earlier about being questioned about whether I was going to use the term *climate change* in my talk, had I acquiesced and actually not used this term, then the implication is that my values in that instance did not include, or at least did not prioritize accuracy and honesty. Similarly, that the world's largest CO_2-emitting countries have failed to meet their emissions targets in the Paris Agreement is an indication that values of and around interconnectedness and interdependency are not consistently being prioritized. Now, part of the rub here probably has to do with the fact that we, both individually and collectively, have many values that can sometimes seem and be at odds with one another. For example, in my world of migration research, discussions and debates about the volume and impacts of immigration to the United States often favor economic arguments and priorities at the expense of other perspectives (National Academies of Sciences, Engineering, and Medicine 2017; Sturdevant 2018). And, while I can agree or disagree with the outsized role of economics here – and I happen to disagree – at least the underlying values are made explicit.

So, in discussions and debates about the human dimensions of climate and environmental change, what do we and what should we value both individually and collectively? I already mentioned accuracy and honesty in the words and definitions that we use to think and communicate about different types of phenomena like climate and environmental refugees. With respect to our actions, attention and commitment to the interconnections, interdependencies, and shared fates of people, populations, and places is also important. Of course, if we zoom out from here just a bit, all of these seem to boil down to three *value baskets*. The first basket includes the dignity and worth of oneself and of others. The second basket includes individual and collective responsibility. And the third basket includes sympathy, empathy, and justice (Decety and Cowell 2015).

Throughout this volume, contributors explore the potential for replacing values of domination and exploitation – made normal through ideological practices like settler colonialism, patriarchy, and racism. My colleagues have proposed celebrating an alternative, often marginalized, shared set of values in line with that which I propose here. For example, David Osborn, in Chapter 2, discusses the need to reject settler-colonial ideas and reimagine our relationship with earth – to become "earthbound." Christina Ergas, in Chapter 3, suggests rejecting patriarchal cultures of domination and replacing them with a "culture of care" focused on regeneration, reciprocity, and equity.

Just as a three-legged stool never rocks, the three value baskets discussed earlier offer a firm foundation for approaching and wrestling with the diverse human dimensions of climate and environmental change. The next step, then, is to put them into practice in particular ways. In the following sections, I offer one such framework for this and a corresponding example on the ground.

ACT now – the Chumash Heritage National Marine Sanctuary

The other day, I was talking with a friend of mine who, like many people, struggled quite a bit with their mental health during the COVID-19 pandemic (Yu 2022). Among the many things we discussed during this conversation were different approaches to mental healthcare. One specific approach that we discussed is Acceptance and Commitment Therapy (ACT). ACT involves two basic and largely sequential steps (Hayes 2005). First, a person has to *accept* what they have. That is, they have to accept the hand that they've been dealt in life and all of the negatives and positives that come with it. Second, they have to *commit* to a course of action that is consistent with who they are and who they want to be at a fundamental level. Importantly, before a person can commit to a course of action, they have to pause between these two steps and think deeply about and decide what they truly value. Absent this, they are like a rudderless ship.

The ACT model serves as sort of a psychological tool that we can use on a global scale and must rely on the use of accurate words, intentional actions, and a new dominant shared value system. Taken in the context of climate change, this means we need to face head-on the dire consequences of climate inaction. We need to call climate change what it is, a crisis, an emergency, a disruption. We need to call climate migrants what they are, refugees. We need to turn toward, instead of away, the hardships disproportionately impacting the world's poorest nations. Only once we have faced reality can we clearly commit to action – action driven by our values and expressed through our words and deeds.

Listening to the news the other day about efforts by the Chumash Tribe to create and manage a proposed new marine sanctuary called the Chumash Heritage National Marine Sanctuary along California's central coast (Sommer 2023), it occurred to me that the tribe is employing the ACT model in its efforts. This can serve as a guide for what ACTing on climate change looks like in practice.

Recognizing the concurrent crises of climate change and Indigenous erasure, the Chumash Tribe proposed the marine sanctuary. In doing so they plan to protect 156 miles of California coastline from oil drilling and other environmental threats. They also seek to provide 600 jobs to the local community and $23 million in economic activity (Chumash Heritage National Marine Sanctuary 2023).

If the first step involves accepting, both individually and collectively, the scary realities, complexities, and uncertainties of where currently we find ourselves, which is a world barreling toward climate and environmental catastrophe, the Chumash Tribe have taken this step by, first and foremost, calling attention to the historical dispossession of Chumash lands and waters. They directly and clearly identify the threat to their people and the planet, refusing to mince words.

If the second step involves committing to a course of action that is consistent with who we are and want to be, again both individually and collectively, the Chumash have also been clear and resolute. They saw a crisis and created a solution. In the process, they did not accept the status quo but forced a new path consistent with their values. For instance, one potential sticking point in the development of the sanctuary was whether and to what extent the Office of National Marine Sanctuaries in the National Oceanic and Atmospheric Administration (NOAA) would manage the sanctuary. However, as Violet Sage Walker, Chairwoman of the Northern Chumash Tribal Council, has made clear, "[w]e are not wanting to be employees of NOAA. We are wanting to be separate and equal so that we have autonomous decision making."

Earlier, I suggested that three value baskets might help to inform and guide courses of action. To me, it seems that the Chumash Tribe clearly value the dignity and worth of themselves and others, the individual and collective responsibility that comes with this, and justice in both process and goal. Of course, whether based on these values and/or some other set, I hope it is clear that it is time to ACT now to tackle the climate crisis before it is too late.

Conclusion

The Chumash Tribe's efforts to propose and manage a new marine sanctuary along California's central coast demonstrate how we, as people who care about climate justice, can ACT now. In their work, they use their words and actions to support a set of shared values. As people who seek a resolution to the climate crisis, we need to be honest and clear about the problems we face. This involves using accurate language when we discuss the climate crisis and clear discussion of the consequences of inaction. We need to accept the dire reality. Then we need to mobilize and commit to a course of action in line with our shared values of equity, fairness, and justice.

References

Bearak, Max and Nadja Popovich. 2022. "The World Is Falling Short of Its Climate Goals. Four Big Emitters Show Why." *The New York Times*. Accessed December 13, 2023. www.nytimes.com/interactive/2022/11/08/climate/cop27-emissions-country-compare.html

Chumash Heritage National Marine Sanctuary. 2023. Accessed November 10, 2023. https://chumashsanctuary.org

Curtis, Katherine J., Elizabeth Fussell and Jack DeWaard J. 2015. "Recovery Migration After Hurricanes Katrina and Rita: Spatial Concentration and Intensification in the Migration System." *Demography* 52(4): 1269–1293.

Decety, Jean and Jason M. Cowell. 2015. "Empathy, Justice, and Moral Behavior." *AJOB Neuroscience* 6(3): 3–14.

DeWaard, Jack. 2020. "Hurricanes and Other Extreme Weather Disasters Prompt Some People to Move and Trap Others in Place." *The Conversation*. Accessed December 13, 2023. https://theconversation.com/hurricanes-and-other-extreme-weather-disasters-prompt-some-people-to-move-and-trap-others-in-place-129128

DeWaard, Jack, Katherine J. Curtis and Elizabeth Fussell. 2016. "Population Recovery in New Orleans After Hurricane Katrina: Exploring the Potential Role of Stage Migration in Migration Systems." *Population and Environment* 37(4): 449–463.

El-Hinnawi, Essam. 1985. *Environmental Refugees*. Nairobi: United Nations Climate Programme.

Fussell, Elizabeth, Katherine J. Curtis and Jack DeWaard. 2014. "Recovery Migration to the City of New Orleans After Hurricane Katrina: A Migration Systems Approach." *Population and Environment* 35: 305–322.

Gabour, Jim. 2015. "A Katrina Survivor's Tale: 'They Forgot Us and That's When Things Started to Get Bad.'" *The Guardian*. Accessed December 13, 2023. www.theguardian.com/us-news/2015/aug/27/katrina-survivors-tale-they-up-and-forgot-us

Gemenne, François. 2015. "One Good Reason to Speak of 'Climate Refugees.'" *Forced Migration Review* 49: 70–71.

Hayes, Steven C. 2005. *Get Out of Your Mind and Into Your Life: The New Acceptance and Commitment Therapy*. Oakland: New Harbinger Publications.

Hersher, Rebecca. 2017. "Climate Scientists Watch Their Words, Hoping to Stave Off Funding Cuts." *National Public Radio*. Accessed December 13, 2023. www.npr. org/sections/thetwo-way/2017/11/29/564043596/climate-scientists-watch-their-words-hoping-to-stave-off-funding-cuts

Hickel, Jason. 2020. "Quantifying National Responsibility for Climate Breakdown: An Equality-Based Attribution Approach for Carbon Dioxide Emissions in Excess of the Planetary Boundary." *The Lancet Planetary Health* 4(9): e399-e404.

International Organization for Migration. 2014. *IOM Outlook on Migration, Environment, and Climate Change*. Geneva: International Organization for Migration.

Millman, Oliver. 2022. " 'No Safe Place': Kiribati Seeks Donors to Raise Islands from Encroaching Seas." *The Guardian*. Accessed December 13, 2023. https://www.theguardian. com/environment/2022/nov/18/cop27-kiribati-donors-raise-islands-sea-level-rise

National Academies of Sciences, Engineering, and Medicine. 2017. *The Economic and Fiscal Consequences of Immigration*. Washington, DC: National Academies of Sciences, Engineering, and Medicine.

National Aeronautics and Space Administration. 2022. *Scientific Consensus: Earth's Climate Is Warming*. Washington, DC: National Aeronautics and Space Administration.

National Centers for Environmental Information. 2021. *Billion-Dollar Weather and Climate Disasters*. Washington, DC: National Oceanic and Atmospheric Administration.

Plumer, Brad, Lisa Friedman, Max Bearak and Jenny Gross. 2022. "In a First, Rich Countries Agree to Pay for Climate Damages to Poor Nations." *The New York Times*. Accessed December 13, 2023. www.nytimes.com/2022/11/19/climate/un-climate-damage-cop27. html

Ripple, William J., Christopher Wolf, Thomas M. Newsome, Phoebe Barnard and William R. Moomaw. 2020. "World Scientists' Warning of a Climate Emergency." *BioScience* 70(1): 8–12.

Shepherd, Christian, Emily Rauhala and Chris Mooney. 2023. "As the World Sizzles, China Says It Will Deal with Climate Its Own Way." *Washington Post*. Accessed December 13, 2023. www.washingtonpost.com/climate-environment/2023/07/19/climate-change-heat-wave-china/

Sommer, Lauren. 2023. "After Decades, a Tribe's Vision for a New Marine Sanctuary Could Be Coming True." *National Public Radio* August 10, 2023.

Sturdevant, Lori. 2018. "Immigration Has a Business Case in Minnesota – And a Faith Case." *Star Tribune*. Accessed December 13, 2023. www.startribune.com/immigration-has-a-business-case-in-minnesota-and-a-faith-case/469059683/

US Department of the Interior. 2022. "Biden-Harris Administration Makes $135 Million Commitment to Support Relocation of Tribal Communities Affected by Climate Change." Press release. November 30, 2022.

United Nations General Assembly. 1951. *United Nations Convention Relating to the Status of Refugees*. Geneva: United Nations High Commissioner on Refugees.

Yu, Verna. 2022. " 'What Was It For?' The Mental Toll of China's Three Years in COVID Lockdowns." *The Guardian*. Accessed December 13, 2023. www.theguardian.com/world/2022/dec/17/what-was-it-for-the-mental-toll-of-chinas-three-years-in-covid-lockdowns

PART 4

Amplifying stories on the margins

11

DISMANTLING THE SETTLER PARADIGM IN INDIGENOUS CLIMATE RESILIENCE

Aiyana James and Laura Laumatia

The problem: Climate change poses disproportionate threats to tribal communities, given their unique land relationship and social and economic vulnerabilities. Climate change impacts are worsened by the history of colonial extraction that has already inflicted catastrophic environmental and cultural damages on Indigenous people. Contemporary climate solutions are at risk of perpetuating settler-colonial approaches and exacerbating historical harms.

The solution: Effective partnerships and collaborations with tribal communities entail understanding each tribe's unique history and how they are impacted by past and current policies that have taken tribal resources. Real commitment to supporting tribes restoring their lands and building community resilience is a long-term effort that requires confronting painful histories and uncomfortable realities about settler dispossession of tribes from their lands, culture, and knowledge systems. Would-be allies in climate work must be prepared to interrogate their own role in these systems if they hope to be effective in dismantling these inequitable systems.

Where this has worked: Across the country, tribal communities are demanding equitable solutions to the climate crisis and demonstrating leadership in taking transformative action. Effective partners in this work have recognized how centering tribal leadership and self-determination, as well as recognizing historic disparities, can lead to powerful climate action.

DOI: 10.4324/9781003437345-16

Over the course of 2023, we have heard the clarion call: act now, as the window of opportunity to rein in warming temperatures is rapidly closing. We've witnessed a devastating year of climate-exacerbated disasters across the globe. In the United States and Canada, these disasters have disproportionately impacted Indigenous communities. In Canada, according to the Canadian Interagency Forest Fire Centre, approximately 18.5 million hectares (nearly 46 million acres) have burned in a devastating fire season (Canadian Interagency Forest Fire Center, Inc. 2023). Though First Nations people make up about 5% of Canada's total population, more than 42% of wildfire evacuations from July of this summer were of predominantly First Nations communities (Webber and Berger 2023). In August, the world watched aghast as on the island of Maui, the historic city of Lahaina was burned to the ground, killing 99 people. Native Hawaiians and global Indigenous activists have since repeatedly detailed how changes to native forest and plant species, dismantling of traditional Hawaiian land management practices, and ongoing settler development patterns contributed to this event (Arango 2023; Bonilla 2023).

Here, where we come together as two climate change professionals, one Indigenous (Aiyana), one White settler (Laura), working together to project threats and plan for resilience in our tribal community in northern Idaho, a sense of urgency presses us to work harder, faster, better to protect what remains precious to our community. Yet we often feel we are in a constant state of emotional whiplash. Globally, we hear world leaders speak to the critical need to center Indigenous communities in adaptation planning, acknowledging how Indigenous land relationships have sustained biodiversity and protected forestlands far more successfully than non-Indigenous jurisdictions (United Nations Climate Change News 2022). The US federal government, as part of its climate action agenda, has focused on tribes and environmental justice through its pledge that 40% of federal investments in climate strategies should go to marginalized and disadvantaged communities (Executive Order No. 14008, 2021). Nevertheless, in our daily work, we find our challenges are not simply the ecological threats of climate change, but the hidden structures of settler colonialism that continue to permeate existing domestic legislation and enable the continued extraction of Indigenous resources, both physical and sociocultural.

Effects of settler colonialism on tribal sovereignty and ecosystem stability

In her 2019 book, *As Long as Grass Grows: The Indigenous Fight for Environmental Justice, from Colonization to Standing Rock*, Indigenous scholar Dina Gilio-Whitaker argues that for Indigenous environmental justice to be realized, we must adopt a new framework that explicitly recognizes the ongoing shackles of settler-colonial history with regard to Native nations *and* centers environmental justice from an Indigenous self-determination perspective. We offer here our observations of how this history and ongoing settler-colonial structure impacts our work, and how it demands frank and often uncomfortable conversations about how the pervasiveness of settler entitlement to property is an obstacle to the transformative work needed to address the crisis before us.

There are 574 federally recognized tribes in the United States, and an additional 63 state-recognized tribes (National Conference of State Legislatures 2016; US Department of Interior 2022). Each has a unique history and relationship with the United States, and most also have a story of how settler encounters have disrupted their land relationship. The community we serve, the Coeur d'Alene Tribe, provides a plain example of how settler colonialism, a paradigm supported by legal concepts rooted in the 15th-century Doctrine of Discovery, is not merely a historical occurrence but an ongoing threat to the tribe's ability to restore and protect the health of its landscapes. Settler colonialism is defined by seminal scholar Patrick Wolfe as, "an ever-incomplete project whereby colonisers repetitively seek to impose and maintain White supremacy" (Wolfe 2006; 2016). The legalization of this ongoing project began in the Americas even before the arrival of Columbus through a series of papal bulls issued by the Catholic Church that codified the rights of Christian people to take the lands of non-Christians, recognizing only a right of occupancy by Indigenous peoples. As this legal doctrine developed, it grew to include a right and responsibility to "civilize" non-Whites, and the crux of what was considered civilization was individual ownership of property. This framework became embedded in US law through a series of 19th-century court decisions and legislation but also became embedded in US beliefs about its own role in the world order, exemplified by the idea of Manifest Destiny, the 19th-century belief that US settlers were destined by God to expand to the Pacific Ocean (Calderon 2014).

Prior to settler arrival, the Coeur d'Alene landscape sustained its people with abundance for millennia. The tribe inhabited a landscape of nearly five million acres of the western Rockies and the Palouse Prairie, with Coeur d'Alene Lake and its tributaries at its heart. This landscape provided all that the Coeur d'Alene people needed: anadromous Chinook salmon and Redband trout populated the southern waters, while the lake and eastern rivers teemed with westslope cutthroat trout and whitefish, ensuring that tribal members had access to fish year-round. Prairies and wet meadowlands provided camas, bitterroot, and a myriad of medicinal plants. Its aboriginal landscape provided all that was needed for the people.

That abundance was an enormous draw for mid-19th century settlers, whose increasing encroachment on Coeur d'Alene lands forced the tribe into several decades of negotiations that saw them cede 90% of their landscape to the United States in order to protect Coeur d'Alene Lake and the surrounding lands that should still have provided them with access to game, fish, berries, roots, and medicines. For a short time, the tribe incorporated new ways of farming and adapted to a changing economy, and it prospered. But settler hunger for Coeur d'Alene land and resources led the United States to break the promises it made to the tribe in 1889 that "the Coeur d'Alene Reservation shall be held forever as Indian land and as homes for the Coeur d'Alene Indians" (US Executive Order 1889). In 1906, the United States forced its allotment policy onto the Coeur d'Alene Tribe and parceled up its land into 160-acre sections. The justification for the allotment process was the assertion that individually held property would accelerate the civilization and prosperity of Native Americans across the nation. In reality, it was a massive settler land grab. Deliberately removing tribal members from lands near Coeur d'Alene Lake and from the forested mountain areas, the government opened up Coeur d'Alene lands for non-Native homesteading.

Passage of the Burke Act, an amendment to the original allotment policy in 1906, enabled unscrupulous Bureau of Indian Affairs agents to collaborate with their acquaintances to quickly move tribally held allotments into "fee" or privately held land status, subjecting it to taxation, at times without the tribal landowner's knowledge. As a result, many tribal members lost their allotments to tax fore-closures, accelerating land loss. By 1933, the Coeur d'Alene Tribe retained less than 20% of the land within its boundaries (Dozier 1962; Watkins 1996). During this same time period, upstream intensive mining for silver and lead in the ceded lands of the Coeur d'Alene River rapidly contaminated downstream tribal waters with mining tailings. Over the next century, more than 75 million tons of heavy metals-laden sediment would come to rest at the bottom of Coeur d'Alene Lake, which remains today in the middle of one of the largest Superfund sites in the country. The contamination already has had widespread impacts on the flora and fauna that have made them unsuitable for tribal subsistence. Projected warming lake temperatures over coming decades heighten the risk of these contaminants remobilizing into the lake's waters, which would threaten the well-being not only of the tribe but the entire region (National Academies of Sciences, Engineering, and Medicine 2022).

Today, the long-term impacts of allotment persist, as the majority of land is still owned by non-Natives, and the predominant land use has been natural resource extraction. Sixty-five percent of historic wetlands on lands of the Coeur d'Alene Tribe have been lost or degraded by land-use conversion. At least 115,000 acres of forestland on the reservation have been cleared for agriculture and development over the last century, and the remaining forests are tremendously pressured, with about three-quarters of commercial timberlands on the reservation losing forest cover in the last two decades (Coeur d'Alene Tribe 2000). Once-abundant fisheries have been extirpated by dams or dramatically reduced by habitat changes, cutting off the ability of tribal members to access their first foods. The tribe has spent decades reacquiring lost lands and invested millions of dollars in restoring degraded ecosystems, but climate change threatens to unravel their efforts, as increasing drought, extreme heat, decreasing snowpack, and intensive precipitation are already evident (Coeur d'Alene Tribe 2023). Land fragmentation from historical policies as well as contemporary land-use policies that prioritize private property rights hamstring tribal ability to restore ecosystem function.

Barriers, choices, and actions for ensuring tribal collective continuance

Into this context we enter, two women of different generations and different backgrounds, both wishing to support the tribe's collective continuance, defined by Indigenous climate scholar Kyle Powys Whyte as the ability of a community to adapt to external forces from naturally occurring environmental change (Whyte 2016). In a community whose collective continuance remains threatened by ongoing extractive practices, we must call out how these structures persist, and are even perpetuated, by seemingly well-meaning agencies, individuals, and collaborators at this unique moment when the United States has allocated unprecedented financial and technical resources for climate adaptation and mitigation (e.g., Inflation Reduction Act passed in 2022). Though we share a sense of urgency to address the climate crisis, much of our experience has elucidated for us how settler colonialism still raises its ugly head in our everyday work. We have witnessed the imposition of federal program guidance that skirts or ignores resource inequities, hampering the ability of tribal communities to even access funding opportunities. We have seen how federal policies quietly undermine tribal sovereignty by requiring tribes to partner with nongovernmental organizations or involve state agencies in the management of tribal lands. We have heard colleagues in regional conservation organizations lament tribal concerns about cultural impacts of renewable energy development as slowing down something that is "too important" for interference, favoring expediency over real environmental justice. Across the country, these organizations are aided by federal agencies that dismiss tribal concerns about energy projects and mining permits, using bureaucratic regulations and timelines to push aside tribal leaders' concerns about lack of consultation (Bryan 2023). One of our most disheartening observations has been how quickly the very agents that have benefited from threatening tribal collective continuance now stand to benefit from tax credits and subsidies. The Inflation Reduction Act has received much criticism for funding corporations for voluntary land-use practices that require no real transformation to land-use practices, may have no impact on reducing carbon emissions, and may even fund pollution. Hughes et al. (2023:501) note how this form of greenwashing is simply another settler structure:

> Even if we think of climate change mitigation and green development as social and environmental justice projects – though often, they are – they are frequently at odds with decolonization, since they do not involve the return of land to Indigenous stewardship or the redress of settler violence.

How, then, do we resist these settler tendencies and re-center those who have already lost the most and who still stand to lose the most? How do we use both our professional positions and our positionality, or our individual context and identities, to challenge settler demands for property, productivity, and consumption that continue to marginalize tribal self-determination and land relationship? We share here our two perspectives on how we must approach this work.

Aiyana's perspective

I identify as many things, a woman, daughter, sister, friend, but the core of my identity is being a member of the Coeur d'Alene Tribe. This identity consists of many traits that have been influenced by my family, my friends, my ancestors, but it is the natural world that brings me closer to who I truly am. As Coeur d'Alene people, we have always lived in coexistence with the land and our nonhuman relatives. Our understanding of the interconnectedness of our spirit with the water in our lakes and rivers, the trees in our forests, or the animals throughout our ecosystem, is incorporated and upheld in all of our doings as a people. Never has there been any part of the earth that has been viewed as something outside of our own identity, but rather seen as an extension of our own self. We hold great value in our relationship with our natural environment, to the extent that our entire being is reliant on how well we protect and prioritize the health of the earth. The core of our existence is based on this inherent principle, and in return we are gifted with protection and care from all of nature's beings. Through acts of settler colonialism and the rise of capitalism, however, dominant society as a whole has not upheld its end of this deal. Climate change and the climate crisis can be traced back to many things, fossil fuels, deforestation, or unsustainable development, but it is ultimately driven by our disconnect from the natural world, and therefore ourselves and our spirit.

Settler colonialism introduced many things to Indigenous people, including disease and land dispossession, but perhaps most destructive was the idea that the way we were living had no value to the colonizer, and that our only value would come if we yielded to cultural assimilation and abandoned our traditional ways. This process of assimilation and cultural genocide spread quickly into US Indian policy, expressed most plainly in the infamous statement that US leaders should "Kill the Indian in him, and save the man" (Pratt 1892). This ideal was represented in nearly every policy imposed on Indigenous people throughout the nation, leading to decades of generational trauma that persists to this day. The removal of our culture and traditions led to a shift in the way that we interacted with our environment. In this assertion of cultural erasure, we were punished for doing things that brought us closer to ourselves, our community, and our land, like speaking our language, holding ceremonies, or gathering and making medicine. This forced attempt of detachment from our traditions, and therefore our identity, severely damaged Indigenous people's connection to the land with which our very being is interconnected. Though many Indigenous communities are still maintaining that relationship despite the constant barriers enforced on us to break that bond, it is becoming harder to preserve due to the lack of accessibility to the resources that helped us to maintain that connection.

As climate change increasingly threatens our survival as humanity, we are witnessing the rise in recognition of traditional ecological knowledge (TEK) as a valuable tool of resilience. TEK is most commonly defined as "bodies of dynamic and experimental knowledges gained over time by Indigenous peoples, often associated with a specific place" (Fifth National Climate Assessment 2023). TEK is not so much a science as it is a philosophy: "[it's] not just about understanding relationships, it is the relationship with Creation. TEK is something one does" (McGregor 2008, as cited in Whyte 2013). Though this philosophy can be seen throughout Indigenous history, rather than being labeled as something such as TEK, it was simply explained as the way things ought to be, or the way we lived our life in relation to our land.

The growth in recognition of this ideology by non-Indigenous peoples has shifted the way broader society thinks about solving the climate crisis. It is forcing people to recognize that the use of Indigenous knowledge is crucial to ensuring the health and well-being of the land and its inhabitants. As someone who is passionate about solving climate change using the ideals held in Indigenous philosophies, I realize that this could finally be the answer to restoring balance in our societal coexistence with our natural world. However, I still have my doubts about the level of integrity behind this acknowledgment by non-Indigenous peoples. The intuitive relationship with our environment that led to our survival and abundance as Indigenous people, once ridiculed by the oppressor, is at the forefront of the values now touted in climate resilience ideologies of modern society, and even encouraged by inherently colonialist structures. I am skeptical of these claims by western government systems that claim to champion the very beliefs they spent centuries trying to destroy.

I am hardly alone among Indigenous people in my distrust. It is easy to comprehend the rational fear of betrayal that there are not sincere and ethical motives behind this integration of TEK into federal and agency actions and philosophies. The hesitation stems from the irony that we are asked to trust these efforts by colonialist structures without their accepting fault for the erasure of the very way of thinking that is integrated within the culture of Indigenous people. The paradox of moving from the forced detachment and belittlement of these values that are representative of Indigenous identity, to now promoting them as the solution to a problem created at the hands of settler colonialism is clear. I do believe the solution to the climate problem is restoring the values that have sustained us since time immemorial, yet, our society has to move past the Lockean paradigm that sees Indigenous people as wasting our land and not using its resources correctly (Locke 1690). The distrust that is attached to this ironic "discovery" of TEK ideology is driven by my fear of an attempt by government systems to gloss over their history of oppression and invalidation of Indigenous environmental values and relationships. It is easy to see it as an "attempt to relieve the settler of feelings of guilt or responsibility without giving up land or power or privilege, without having to change much at all" (Tuck and Wayne 2012) beside the words that they speak. This is advanced by the continuing efforts by environmental activists to express allyship for Indigenous environmental sovereignty and values in an attempt to separate themselves from the doings of their ancestors. Though it is easy for modern settlers to detach themselves from their colonialist history, there is often the same underlying theme of appropriation in the philosophical explanations of why we should fix our relationship with the natural world.

Within the climate movement, there is recognition that the dominant society's central paradigm of consumption and profit is flawed. Since people too rarely recognize the philosophical underpinnings of their own decision-making, they are seemingly looking for others to present solutions. In environmental philosophy, there has been an influx of "new" theories that aim to explain how we should view the world, yet they seem to be recycling Indigenous philosophies and values without crediting their roots, or the efforts to eliminate them. Ecocentrism, for example, explains a worldview that sees all of nature as having inherent value and is centered on nature rather than on humans (University of New South Wales 2017). Another view is deep ecology, "the belief that humans must radically change their relationship to nature from one that values nature solely for its usefulness to human beings to one that recognizes that nature has inherent value" (Madsen 2023). There is great value in these philosophies. The idea that we should live with the view of nature as having intrinsic value resonates with me. My educational background that has focused on environmental philosophy is in part due to these theories helping to spark my passion. Yet I will not ignore the claiming of environmental leadership by those articulating these theories who persist in refusing to acknowledge their ancient roots or those who refuse to move from philosophy to policy so that Indigenous land relationship can be restored.

As a young professional in the climate arena, I recognize that I have challenges in tackling the power dynamics that would perpetuate settler appropriation of Indigenous philosophies, taking on Indigenous knowledge systems as property. However, I also feel that it is my responsibility to reclaim my Indigenous voice at the center of these philosophies. Though I am entitled to my distrust of suspicious actors, I do recognize the significant change that can be made from the inclusion of Indigenous philosophies in climate change solutions. I would ask that Indigenous people not be responsible for the weight of change that needs to be accomplished, since we have had little cause in this issue, but that we can be a guide to how we must coexist with each other and our natural world. We hold a great history of knowledge and experience that represents the success of our ways of interacting and being as one with the environment. This knowledge must be recognized, credited, and accepted by our colonial systems if we are to successfully address this climate crisis we are in.

Laura's perspective

I am a White settler who has worked in Indigenous communities in areas of natural resources, environment, and community development for over 25 years. As a settler now working for a tribal nation in the climate resilience effort, I must be cognizant of the messiness of my role. Even as I express my sense of urgency to plan and adapt on behalf of the community I work for, I must be sensitive to Kyle Powys Whyte's critique of non-Indigenous climate activists that use projections of climate disasters that mirror what Indigenous communities have already experienced in terms of environmental, social, and cultural loss (2018). Whyte describes how environmental and physical settler-colonial dispossession of Indigenous communities from their lands has already effectively severed relationships and sense of identity. He warns of past allies who "exploited Indigenous peoples as an effort to boost their own senses of righteousness" (Whyte 2018:232), using their own sense of morality to enact destructive policies like allotment in their desire to decide what was best for tribal communities. He calls out the contemporary danger of White saviorism, of allies positioning themselves to lead the way to environmental salvation. Whyte does not shy from the uncomfortable conversation, calling out those who believe there is a chance for the "right allies to save these remaining Indigenous peoples and to learn from them about how the rest of humanity can save itself" (Whyte 2018:236), engaging in what he calls an "allyship of innocence" (Whyte 2018:237) that is unwilling to dismantle the structures that have led to Indigenous dispossession.

I am privileged to work in a tribal community that is renowned for its environmental leadership, and I am grateful for the opportunities it has afforded me to work with a sense of purpose and solidarity. However, if I am to truly commit to the unsettling of climate action, I must start by honestly facing the uncomfortable truths of my role and be ready to face my responsibilities. McGuire-Adams (2021) details ways that settlers can work to undo the settler-colonial structure by engaging in self-reflection to ensure actions align with words, being accountable for how past actions have caused harm, and accepting that work may be uncomfortable. I must acknowledge that it is not my role to determine the priorities and desired outcomes of the community I serve. But the burden of unsettling in this arena should not fall on Indigenous shoulders, and I have a role to play that goes beyond simply understanding the biophysical impacts of climate change. My work has often required me to publicly represent environmental concerns as part of my outreach responsibilities. To be effective, I must be a student of the political history of my community, be direct with partners and collaborators about how past and current land policies impact tribal sovereignty, and hear the criticisms and concerns of community leaders and colleagues when I fall into the trap of replacing their goals with my own. And I must make the spaces and step aside for young leaders like Aiyana to reclaim and restore their own stories and their own knowledge systems.

As a professional working in climate change, I am inundated daily with messages about the urgency of climate work and recommendations for solutions. It would be all too easy to succumb to the temptation to move forward with my own ideas for the sake of expeditiousness. Resisting this temptation requires ensuring that I respect the processes and protocols that tribal leadership has established to ensure that they and community members have an opportunity to review and critique any plans and projects that I may develop. In our recent work on assembling climate data and projections for the community, this entailed sending each chapter draft out to multiple reviewers within the government and tribal entities, meeting with key staff one-on-one, and presenting drafts to internal tribal member committees for their input and editorial suggestions several times before advancing the final draft for approval. This process was time consuming, and likely added several months to the completion of the document, which may have only taken a fraction of the time if we had outsourced the effort to a consultant. However, the time spent on this effort contributed to rich conversations about individual concerns and deepened my own understanding of the breadth of impacts of climate change and how it uniquely affects the tribal community, including how these impacts are extensions of damages already done by colonialist extraction. As a result, the document itself is better, key community leaders more broadly understand our climate work, and we are better poised to develop plans that are aligned with tribal values. Additionally, the report, which attempts to integrate past environmental impacts, as well as projected changes, provides a mechanism for outside collaborators to access information about the unique Coeur d'Alene context as well.

Conclusion

In the midst of our writing process, the World Meteorological Organization reports that 2023 is ending as the hottest yet in human history (Dance 2023). World Meteorological Organization Secretary-General Petteri Taalas describes the ever-increasing temperatures, sea-level rise, and greenhouse gas levels as "a deafening cacophony of broken records." For our planet to remain inhabitable for humanity, we need a transformation, not just in our environmental practices but in the way we tell the story of how and why this climate crisis has occurred. We did not arrive at this environmental catastrophe overnight. The Coeur d'Alene experience is illustrative of how climate change is the culmination of centuries of disruption to the human relationship to the rest of the natural world that have created allowances for pollution and environmental degradation for profit while ignoring the concomitant damages to countless Indigenous communities. As climate professionals, Indigenous and settler, we must accept our responsibilities to each other, to our community, and to the earth to dismantle the settler structures that have already resulted in this ecological violence. If we are to move beyond superficial pledges to environmental justice to meaningful centering of Indigenous communities' role in building a collective path forward to resilience, it requires that we address the political and social systems that are embedded in centuries of colonization. We must take an honest look at how environmentalism does not equate to solidarity but can perpetuate settler structures that continue to co-opt and marginalize Indigenous land, knowledge, beliefs, and voices.

This chapter was prepared by Aiyana James and Laura Laumatia in their personal capacity. The perspectives expressed here are solely those of the authors and do not reflect the views or policies of the Coeur d'Alene Tribe.

References

Arango, Tim. 2023. "A Plea from Native Hawaiians: The Future of Maui Rests on Honoring Its Past." *The New York Times*. Retrieved December 1, 2023. www.nytimes.com/2023/10/20/us/maui-lahaina-fires-rebuilding.html

Bonilla, Yarimar. 2023. "A Legacy of Colonialism Set the Stage for the Maui Wildfires." *International New York Times*, p. NA. Retrieved December 1, 2023. https://link.gale.com/apps/doc/A762550760/AONE?u=naal_talladega&sid=googleScholar&xid=055a6f30

Bryan, Susan Montoya. 2023. "Work Resumes on $10B Renewable Energy Transmission Project Despite Tribal Objections." *AP News*. Retrieved December 1, 2023. https://apnews.com/article/sunzia-renewable-energy-native-american-tribes-fadb5ff59237a1e-ca6eed85a999fddb1

Calderon, Dolores. 2014. "Speaking Back to Manifest Destinies: A Land Education-Based Approach to Critical Curriculum Inquiry." *Environmental Education Research* 20(1): 24–36.

Canadian Interagency Forest Fire Center, Inc. 2023. "Wildfire Graphs." Retrieved December 1, 2023. https://ciffc.net/statistics

Coeur d'Alene Tribe. 2000. *EAP Assessment of Environmental Concerns on and near the Coeur d'Alene Reservation*. Volume 2, Part I. Coeur d'Alene, Idaho: Coeur d'Alene Tribe.

Coeur d'Alene Tribe. 2023. *Climate Change on the Coeur d'Alene Landscape: Coeur d'Alene Tribe Climate Impact Assessment*. Coeur d'Alene, Idaho: Coeur d'Alene Tribe.

Dance, Scott. 2023. "This Year Will Be Earth's Hottest in Human History, Report Confirms." *The Washington Post*. Retrieved December 1, 2023. www.washingtonpost.com/weather/2023/11/30/earth-hottest-year-wmo/

Dozier, Jack. 1962. "The Coeur d'Alene Land Rush 1909–10." *The Pacific Northwest Quarterly*, 53(4): 145–150.

Executive order No 14008, 86 Fed. Reg.19.2021.

Gilio-Whitaker, Dina. 2019. *As Long as Grass Grows: The Indigenous Fight for Environmental Justice, from Colonization to Standing Rock*. Boston, MA: Beacon Press.

Hughes, Sara Salazar, Stepha Velednitsky and Amelia Arden Green. 2023. "Greenwashing in Palestine/Israel: Settler olonialism and Environmental Injustice in the Age of Climate Catastrophe." *Environment and Planning E: Nature and Space* 6(1): 495–513.

Locke, John. 1690. Second Treatise of Government. *The Project Gutenberg eBook of Second Treatise of Government, by John Locke*, 2003. Retrieved December 1, 2023. www.gutenberg.org/files/7370/7370-h/7370-h.htm

Madsen, Peter. 2023. "Deep Ecology." *Encyclopædia Britannica*. Retrieved December 1, 2023. www.britannica.com/topic/deep-ecology

McGuire-Adams, Tricia. 2021. "Settler Allies Are Made, not Self-Proclaimed: Unsettling Conversations for Non-Indigenous Researchers and Educators Involved in Indigenous Health." *Health Education Journal* 80(7): 761–772.

National Academies of Sciences, Engineering, and Medicine. 2022. *The Future of Water Quality in Coeur d'Alene Lake*. Washington, DC: National Academies Press.

National Conference of State Legislatures. 2016. *Brief State Recognition of American Indian Tribes*. www.ncsl.org/quad-caucus/state-recognition-of-american-indian-tribes#:~:text=Brief-,State%20Recognition%20of%20American%20Indian%20Tribes,-Updated%20October%2010

Pratt, Richard. H. 1892. The Advantages of Mingling Indians with Whites. In *Proceedings of the National Conference of Charities and Correction*. Boston, MA: Press of Geo. H. Ellis.

Tuck, Eve and Yang K. Wayne. 2012. "Decolonization Is Not a Metaphor." *Decolonization: Indigeneity, Education & Society* 1(1): 1–40.

United Nations. 2022. "How Indigenous People Enrich Climate Action." Unfccc.int. Retrieved December 1, 2023. https://unfccc.int/news/how-indigenous-peoples-enrich-climate-action#:~:text=Indigenous%20peoples'%20practices&text=Active%20

revitalization%20of%20traditional%20technologies,strategy%20to%20mitigate%20climate%20change

University of New South Wales. 2017. "Why Ecocentrism Is the Key Pathway to Sustainability." UNSW Sites. Retrieved December 1, 2023. www.unsw.edu.au/news/2017/07/why-ecocentrism-is-the-key-pathway-to-sustainability#:~:text=Ecocentrism%20finds%20inherent%20(intrinsic)%20value,but%20there%20are%20related%20worldviews

US Department of Interior. 2022. "Tribal Leaders Directory." Tribal Leaders Directory | Indian Affairs. Retrieved December 1, 2023. www.bia.gov/service/tribal-leaders-directory

US Presidential Executive Order Establishing Coeur d' Alene Reservation Boundaries. 1887. Coeur d' Alene/United States Treaty. 1889. Coeur d' Alene/United States Treaty.

Watkins, M. 1996. *Washington Water Power and the Post Falls Dam Report and Source Materials*. Winthrop, WA: Hart West and Associates.

Webber, Tammy and Noah Berger. 2023. "Canadian Wildfires Hit Indigenous Communities Hard, Threatening Their Land and Culture." *Associated Press News*. Accessed December 12, 2023. https://apnews.com/article/canada-wildfire-indigenous-land-first-nations-impact-3faabbfadfe434d0bd9ecafb8770afce

Whyte, Kyle. 2016. "Indigenous Experience, Environmental Justice and Settler Colonialism." *SSRN* Scholarly Paper ID 2770058. Rochester, NY: Social Science Research Network. http://papers.ssrn.com/abstract=2770058

Whyte, Kyle, Rachel Novak, Matthew B. Laramie, Nicholas G. Bruscato, Dominique M. David-Chavez, Michael J. Dockry, Michael Kotutwa Johnson, Chas E. Jones Jr. and Kelsey Leonard. 2023. *Ch. 16. Tribes and Indigenous Peoples. Fifth National Climate Assessment*. Edited by Crimmins, A.R., C.W. Avery, D.R. Easterling, K.E. Kunkel, B.C. Stewart and T.K. Maycock. Washington, DC: US Global Change Research Program. https://doi.org/10.7930/NCA5.2023.CH16

Whyte, Kyle Powys. 2013. "On the Role of Traditional Knowledge as Collaborative Concept: A Philosophical Study." *Ecological Processes* 2(1): 1–12.

Whyte, Kyle Powys. 2018. "Indigenous Science (Fiction) for the Anthropocene: Ancestral Dystopias and Fantasies of Climate Change Crises." *Environment and Planning E: Nature and Space* 1(1–2): 224–242.

Wolfe, Patrick. 2006. "Settler Colonialism and the Elimination of the Native." *Journal of Genocide Research* 8(4): 387–409.

Wolfe, Patrick. 2016. *Traces of History: Elementary Structures of Race*. New York: Verso Books.

12

CRIMINALIZING CLIMATE CHANGE

Defining and responding to ecocide

Taylor June, Hollie Nyseth Nzitatira, and Nicole Fox

The problem: Climate change poses an urgent threat to our planet, undermining ecosystems, livelihoods, and the well-being of humanity. Traditional responses to environmental damage, such as fines and regulations, are often insufficient and fail to adequately hold responsible parties accountable for their actions.

The solution: Taking a restorative and transformative justice approach, we propose the criminalization of ecocide – the destruction of the natural environment. By establishing ecocide as a crime, we can send a powerful message that harming our planet demands a response. Yet rather than a punitive response, we suggest that ecocide must be met with restorative and transformative approaches that seek to repair harms done and change systematic inequalities while also still holding individuals, governments, and corporations accountable for their actions.

Where this has worked: The successful criminalization of ecocide can be seen in the case of Ecuador. In its 2008 Constitution, Ecuador became the first country to recognize the rights of nature. This legal framework acknowledged the intrinsic value of ecosystems and empowered communities to protect their environment. As a result, Ecuador has witnessed positive outcomes, including increased environmental consciousness, restoration efforts, and the preservation of fragile ecosystems. Moreover, numerous countries have implemented restorative justice practices in the wake of violence and harmful actions (e.g., South Africa's Truth and Reconciliation

DOI: 10.4324/9781003437345-17

> Commission following Apartheid). While each practice has been unique and has come with its own challenges, the implementation of restorative justice practices illustrates that crimes need not be met with punitive measures.

On September 23, 2022, activists marched with signs advocating for the criminalization of ecocide as part of a global movement calling for urgent action. In Manhattan, for instance, people marched from Foley Square to Battery Park bearing signs that read "Stop Ecocide," "Make Ecocide a Crime," and "Climate Justice = Healing." These signs were translated and displayed in marches across the world including but not limited to events in Tokyo, Berlin, and Bengaluru. The momentum behind these activists' calls has grown significantly, driven by the imperative to hold individuals and corporations accountable for their destructive actions against the planet. In fact, many countries, including Ukraine, Russia, Vietnam, and France, have already passed laws criminalizing ecocide, and others, such as Mexico, the Netherlands, Belgium, and Scotland, are considering similar legislation (Kaminski 2023). While the ultimate goal is international recognition of ecocide as a crime, domestic legislation plays a crucial role in supporting and reinforcing this objective. As more national governments concentrate on ecocide, it garners increasing attention and consideration at the international level.

This chapter focuses on these activists' call for action: criminalizing harm to the planet. Specifically, we address the movement to delineate a crime of ecocide, or "extensive damage to, destruction of or loss of ecosystem(s) of a given territory, whether by human agency or by other causes, to such an extent that peaceful enjoyment by the inhabitants of that territory has been severely diminished" (Higgins et al. 2013:257). To date, international environmental crimes have largely pertained to the trafficking of flora and fauna, poaching of animals, and destruction of forests. Yet, climate change, too, causes extreme damage and hardship to the earth as well as to human and nonhuman animals. What is more, climate change may contribute to increased violence, including but not limited to war, genocide, and even intimate partner violence.

We argue that climate change should be criminalized as ecocide in international law, due in large part to the benefits of defining harmful actions (or lack of action) as a crime.[1] Criminalizing climate change and related environmental damage is not simple or straightforward, and care must be taken to ensure criminalization does not result in unintended consequences and disproportionate impacts, particularly on marginalized individuals and communities.

Additionally, while we urge criminalization of climate change, we do not advocate for punitive responses, such as prison sentences. Instead, we join a chorus of activists, organizations, and scholars who argue that ecocide should be met with restorative justice and transformative justice responses. Generally, these approaches emphasize reducing harm, promoting justice, preventing future devastation, and supporting the restoration of environmental and community damages.

In what follows, we begin by outlining some of the major environmental harms that have, to date, been criminalized in international law. Then, we highlight why many activists, scholars, and others argue that climate change should be criminalized as well. Next, we address the benefits and pitfalls of criminalization. Finally, we outline possible responses to these crimes with an emphasis on restorative and transformative approaches.

The road to recognizing ecocide

Crimes are socially constructed actions or omissions that violate established laws. They change with society's perception of the crime and social relations.[2] While laypeople often think of crimes as actions that harm people (e.g., murder, burglary), many other actions have been criminalized, like arson or motor vehicle theft. In recent times, crimes have also encompassed harm to nonhuman animals and the natural environment, known as environmental crimes.[3]

The first laws defining environmental crime were passed by national governments. For instance, the Rivers and Harbors Act of 1899 is typically seen as the first law pertaining to environmental crime in the United States, as it deemed discharging garbage into navigable water (without a permit) as a misdemeanor. Another example can be found in Thailand's 1927 Forest Act. This act aimed to regulate logging activities and protect forests from environmental degradation, in part by establishing a system for forest management and granting power to forest officers to enforce regulations and impose penalties for violations. Under the Forest Act, individuals or companies found guilty of illegal logging or damaging forest areas could face fines and imprisonment.

With the rise of globalization and international organizations like the United Nations, environmental crimes have also become part of international law, or sets of rules, norms, and standards that regulate how countries interact with one another and how they treat their citizens, among other factors. Crimes of international law are set forth in international treaties, or written agreements between states that set forth legally binding standards (Vienna Convention, Article 2 (1)(a)). Treaties regarding environmental crime emerged as environmentally related misconduct was increasingly crossing borders. For instance, as issues like the trafficking of elephant and rhino ivory became more prevalent, governments and the United Nations began to create treaties regarding the trading and smuggling of plants, animals, resources, and pollutants.

Most notably, between 1973 and 1975, countries ratified the Convention on International Trade in Endangered Species of Wild Fauna and Flora. This treaty, which has been ratified by 184 parties as of 2023, regulates the flora and fauna that can cross international borders. A decade later, the 1989 Basel Convention addressed the transfer of hazardous waste between nations as well, seeking to ensure that hazardous waste is disposed of in an environmentally safe manner.

The 2000 Convention Against Transnational Organized Crime took these and related treaties even further by directly addressing transnational crime, or crime that crosses borders. Put another way, prior treaties had set forth regulations regarding the flora, fauna, and waste that could cross borders, though they did not deem actions breaking the regulations as crimes. The Convention Against Transitional Organized Crime, however, directly invokes criminalization. In fact, in the preamble, the drafters of this treaty explicate that their goal was to create an

> effective tool and the necessary legal framework for international cooperation in combating, inter alia, such criminal activities as money-laundering, corruption, *illicit trafficking in endangered species of wild flora and fauna*, offences against cultural heritage and the growing links between transnational organized crime and terrorist crimes. (emphasis added, Preamble 2000)

While the Convention did not explicitly criminalize the trafficking of flora and fauna, it did set the stage for such actions to be treated as crimes given its emphasis on crimes that cross borders.

Today, environmental crime is well established globally, and the International Criminal Police Organization (INTERPOL) regularly publishes notices on individuals who violate international environmental law. For instance, their "red notices" – or alerts that seek the location and arrest of individuals wanted for grave crimes – often include the names of people wanted for trafficking lumber or ivory. In fact, since 1992, INTERPOL has had an environmental crime committee that supports its efforts tied to international crime via specialized working groups in fisheries, wildlife, forestry, and pollution crimes. In 2000, the United Nations also started their own environmental governance project called Wildlife Enforcement Monitoring System, which is aimed to systematically record incidents of poaching and seizures.

While many of these efforts pertain to crimes that cross borders or directly target endangered flora or fauna, international environmental law also sets forth laws concerning natural resources and pollution. Examples of pollution crimes include polluting oceans, trafficking chemicals, and engaging in illegal mining (when the mining processes uses chemicals and/or harmful machinery). While there are unfortunately many instances of pollution crimes, the 2010 BP *Deepwater Horizon* oil spill remains one of the most infamous pollution crimes to date. The oil spill resulted in extensive environmental damage due to the release of millions of barrels of oil into the Gulf of Mexico over several months, ultimately harming the ecosystem, aquatic life, wildlife, and water quality. The oil spill also led to an increase in illegal activity, such as looting and fraud, as individuals attempted to capitalize on the catastrophe. Aside from the extensive cleanup efforts for the oil spill, BP faced criminal charges, and a $20 billion compensation agreement was reached through the Department of Justice (Department of Justice 2015).

While pollution crimes like the BP oil spill are often recognized as crimes, pollution and related environmental harms that contribute to climate change are yet to be criminalized. Nevertheless, actions (or inactions) tied to climate change also result in extreme harm that merits criminalization. Directly, climate change inflicts profound damage upon ecosystems and jeopardizes biodiversity. Elevated global temperatures, altered precipitation patterns, and escalating occurrences of extreme weather events disrupt fragile ecological equilibriums, resulting in habitat loss and species demise. In fact, over one-third of earth's animal and plant species are projected to be extinct by 2050 if current greenhouse gas trajectories continue.[4] This ecological degradation not only erodes the inherent value of biodiversity but also impairs air, water, and sustenance, thus impinging on human well-being and the livelihoods contingent upon thriving ecosystems.

Many researchers have also found that climate change is directly harming certain cultures by infringing on their ways of life (Crook et al. 2018; Crook and Short 2014). In fact, Crook and Short (2014) suggest that "ecologically-induced genocide" is occurring due to the overexploitation of land and water, which in turn results in the destruction of connections between people, culture, and land – connections that are particularly valuable for many Indigenous populations worldwide. As such, they argue that climate change can have genocidal consequences for Indigenous peoples who depend on the land spiritually and physically. Indeed, researchers who have proposed criminalizing ecocide have long linked this crime to the crime of genocide (see overview in Higgins et al. 2013).

More indirectly, climate change can serve as a catalyst for social and political upheaval, fueling diverse manifestations of violence. As global temperatures surge, regions grappling with drought, declining agricultural yields, and/or heightened occurrences of extreme weather events confront amplified competition for scant resources. Such resource-driven conflicts precipitate social turmoil, displacement, and, in extreme cases, armed conflict, engendering contexts of violence and human rights violations.

Climate change additionally intersects with existing social disparities, amplifying vulnerabilities and disproportionately impacting marginalized populations. It often entrenches socioeconomic inequalities, as those with limited means or who are already grappling with socioeconomic challenges typically lack the capacity to adapt to or recuperate from climate-triggered consequences. Such disparities potentiate social unrest and deepen social cleavages within societies, potentially precipitating civil discord, forced displacement, and unprecedented migratory flows, thereby accentuating the precursors for conflict. Climate change intersects with other forms of violence as well, including gender-based violence and human trafficking. In numerous contexts, struggles for resources, rising tensions, and the displacement induced by climate change augment the perils of violence, exploitation, and human rights transgressions, rendering women and marginalized groups particularly susceptible. Moreover, climate-induced disasters often disproportionately impact women, who confront heightened risks of sexual violence, domestic abuse, increased inequality, and exploitation in their aftermath (Annecke 2002; Agwu and Okhimamwe 2009; Rahman 2013).

For these and many other reasons, activists, researchers, scholars, and numerous others have called for climate change to be criminalized as ecocide, or criminalized human activity that violates the principles of sustainability and respect for human rights that causes extensive damage, destruction, or loss to the earth's ecosystems. Those advocating for ecocide often see themselves as global citizens and suggest that there should be an international treaty outlawing ecocide. To date, there are no international laws pertaining to ecocide in times of relative peace, though the Rome Statute (which created the International Criminal Court) does deem "long-term and severe damage to the natural environment which would be clearly excessive in relation to the concrete and direct overall military advantage anticipated" as a war crime (Article 8biv). Moreover, during the last few decades, countries have bound themselves to emission rates in treaties, such as the well-known Kyoto Protocol or the Paris Agreement. However, while these agreements are legally binding to those who ratify them, they do not criminalize breaking emission regulations or any other form of climate injustice committed in or by states who do not ratify such agreements. As such, it is time to consider creating an international convention on ecocide that directly criminalizes major actions resulting in climate change, and in the next section, we outline some of the benefits (as well as the pitfalls) of criminalization.

Why criminalize?

There are numerous reasons behind activists' and scholars' calls to criminalize climate change by creating a convention on ecocide. At the most fundamental level, criminalization ensures that the rules that govern society are outlined in written format, establishing common norms and a shared understanding of a common goal to protect the earth, nonhuman species, and the next generation. People are thus able to better understand what behaviors are expected of them, as well as what consequences they would face for deviating from these acceptable standards.

To be certain, law is not always clearly understood, as complexities with legal language and procedures, inadequate legal education and literacy, socioeconomic barriers, and other disparities inhibit universal understanding. Moreover, law often reflects the views of those in power, and it is well established that laws can also be harmful, especially to minoritized communities. However, laws can also be used to uphold principles of justice and created via an approach that values consensus, equality, and understanding. Accordingly, clearly written laws that outline crimes can serve as a crucial foundation for a just society by establishing comprehensible rules and standards of behavior, promoting consistency and predictability in legal decision-making, and providing a framework for resolving disputes. Laws concerning ecocide can play a crucial role in safeguarding the rights and well-being of entities such as animals, plants, the atmosphere, and the earth that lack representation in the legislative process.

Criminalization via the law can be particularly important for crimes that traditionally go unnoticed and unpunished, as criminalization can change public opinion. For example, for many years, domestic violence was thought of as a private issue, one to be dealt with only in the home. Now, through criminalization and associated public campaigns, domestic violence has been framed as a public issue (Bassadien and Hochfeld 2005; Bui and Morash 2008; Mann and Takyi 2009), underscoring how criminalization can draw attention to societal problems that may have previously been unrecognized or underrecognized. Additionally, acts that are formally deemed crimes tend to receive media attention, in turn enhancing public awareness of the harmful and negative impact of such behavior on society.

In line with acknowledgment of certain acts as publicly harmful and problematic, criminalization can play a significant role in recognizing what has become known as *slow violence*. Coined by Rob Nixon (2011), slow violence refers to a form of violence that unfolds gradually and often imperceptibly, with long-term and far-reaching consequences for human and nonhuman life and ecosystems. Examples of slow violence include environmental degradation, climate change, the slow poisoning of communities through toxic waste, and the erosion of social norms and democratic institutions over time. Unlike direct forms of violence, such as physical assault or war, slow violence is often invisible, taking place over months, years, or even decades. It is also often diffuse and systemic, involving multiple actors and institutions who contribute to the harm but are not immediately responsible for it.

Despite its pervasive and harmful nature, slow violence is often not criminalized due to the gradual and incremental nature of the harm and the complexities of causality and accountability. However, criminalization is valuable because it holds people accountable for their actions, even when they are gradual. Specifically, persons who engage in criminalized behavior often must account for their behavior. As we address in our final section, however, such accountability does not need to be tied to punitive sanctions but rather should be tied to more restorative and transformative responses.

Restorative and transformative justice

By urging criminalization, we are not advocating for punitive responses. In fact, the scope of criminal legal systems – whether at the national or even the international level in the International Criminal Court – is limited, and criminal legal systems are often unable to address the root cause of environmental harm.[5] Responding to climate change and environmental degradation requires systemic change and structural reforms that cannot be achieved through punitive measures alone. Moreover, the high cost of criminal justice processes and enforcement can create a burden on taxpayers and divert resources away from prevention and mitigation strategies, such as investment in renewable energy and reforestation efforts.

Punitive responses, such as incarceration, are also an ineffective deterrent mechanism that most often end up harming victimized populations. Indeed, much research has found that prison sentences do not deter people from committing crimes (Apel and Diller 2017; Dölling et al. 2009). Penalties (when used alone) are unfortunately similar, as punishments are often too low to deter corporate entities from engaging in environmentally harmful activities, and those committing crimes may also be able to easily circumvent consequences through legal loopholes or easily pay off any fee with whatever profits they make in the meantime. Additionally, criminal justice responses disproportionately impact marginalized communities – communities that regularly bear the brunt of environmental harm but continually face barriers in accessing justice due to systemic inequalities.

Accordingly, as ecocide becomes criminalized, we suggest that appropriate responses will not be punitive but rather will involve victim-centered approaches that place the most vulnerable and affected populations at the center of the response. A victim-centered approach to addressing environmental crime and climate change recognizes that the most destructive and powerful actors, such as governments and large corporations, are often the main perpetrators of environmental harm and that they must be held accountable for their actions. Yet, rather than only focusing on punishment, such an approach underscores protecting the rights and needs of individuals and communities who are disproportionately impacted by environmental harm due to power imbalances and social inequalities. By prioritizing the voices and concerns of those who are most affected, a victim-centered approach can bring about meaningful change that addresses the root causes of environmental crime and climate change.

Specifically, we suggest that the criminalization of ecocide should be met with a two-pronged approach that includes restorative justice and transformative justice endeavors (Babakhani 2023; Chaffin et al. 2016; Lynch and Long 2022). Restorative justice seeks to repair the harm caused by a specific crime. It places the focus on the victim and their needs, as well as on the offender's responsibility for their actions. Restorative justice responses thus not only hold the offender accountable for their wrongdoing but also allow for the victim to have a say in the judicial process and ensure they receive compensation for their losses. Transformative justice, on the other hand, goes beyond individual incidents and instead seeks to address the systemic issues at play. Such an approach would be necessary because climate change and environmental crimes are often rooted in larger political and economic structures, such that meaningful change requires transforming those structures via sustainable development, policy reform, and community-led initiatives that prioritize environmental protection and social justice.

While there are many mechanisms of restorative justice, reparations may be a particularly important response to ecocide.[6] Reparations involve the act of making amends for a wrongdoing or an injury that was inflicted on an individual or group, often by providing compensation. For instance, in 2015, the Ecuadorian government awarded $9.5 billion to Indigenous communities affected by oil pollution in the Amazon region. This compensation was meant to cover the environmental and social damage caused by oil drilling by Chevron, including but not limited to medical bills for treating illnesses caused by exposure to pollutants, the restoration of degraded ecosystems, and the cost of property damage. Indeed, reparation may also take the form of providing healthcare and education to those who suffer lasting health impacts from environmental disasters that may have impacted generations to come (such as birth defects, mortality rates, or increased cancer rates) (Gensburg et al. 2009).

There have been broader discussions advocating for reparations from the countries in the Global North to the countries in the Global South as well. While nothing concrete has emerged (and likely would not on the part of the world's richest nations), these discussions have drawn attention to the fact that many of the largest polluters are rich enough to protect themselves from the impact of pollution, while some of the poorest nations pollute the least yet feel the impact of those who pollute the most.

Additionally, this form of justice aligns with the United Nations' loss and damage agreement, which concentrates on addressing the consequences of climate change and endeavors to offer compensation and assistance to communities and countries facing the most severe repercussions. The discourse surrounding reparations from the Global North to the Global South elucidates the stark disparity between affluent nations, sheltered from pollution's ramifications, and impoverished nations, disproportionately affected despite their minimal contribution.

Another form of repair for environmental crimes is the restoration of degraded ecosystems. This would entail taking active steps to restore damaged ecosystems to their former state (when possible), such as planting trees in deforested areas, cleaning up polluted water bodies, and supporting the growth of natural habitats. This form of reparation is beneficial because it focuses on restoring the natural environment, which benefits the affected communities and ecosystems in the long term. For example, the Chesapeake Bay Trust has worked with government agencies, local communities, and businesses to implement projects that support the restoration of Chesapeake Bay, such as planting forests and wetlands, improving water quality, and reducing pollution.

In addition to financial compensation and ecosystem restoration, reparations for environmental crimes can also involve institutional reforms, such as strengthening of environmental regulation and enforcement. This is important as it helps prevent similar environmental crimes from occurring in the future. For example, in the aftermath of the BP oil spill in 2010, the US government implemented institutional reforms to strengthen offshore drilling regulations and improve emergency response preparedness.

Of course, the determination of appropriate reparations is a daunting task due to the grave, and sometimes permanent, destruction to humans, entire species, and the planet. Furthermore, power imbalances often put marginalized groups at a disadvantage in ways that limit them from gaining access to justice to start the reparations process. Documenting environmental injustice and garnering evidence can sometimes take decades, and it often necessitates that people in power help move the process forward. Moreover, the procedure of implementing reparations must be both effective and transparent to ensure that all of the involved parties are satisfied, meaning that reparations – no matter the form they take – must be implemented thoughtfully and can take significant time and money to be done well.

Finally, while reparations and other mechanisms of restorative justice may be used to respond to specific incidents of ecocide, the aftermath of environmental crimes demands a comprehensive approach that goes beyond restorative justice. In this case, we suggest that responses to climate injustice must also be met with transformative justice. Transformative justice is a holistic approach to social justice that seeks to change entire societal systems and create sustainable and equitable outcomes. Put simply, proponents of transformative justice suggest that any justice mechanisms must address systematic and structural issues.

To be clear, while restorative justice approaches have been implemented, transformative justice approaches have yet to see implementation. Indeed, large-scale systematic change is incredibly difficult and takes time; and as such, we are not able to provide concrete examples of successful transformative justice endeavors. Nevertheless, to truly confront the climate crisis, criminalization must be accompanied by approaches that seek to address the root causes of ecocide, including unsustainable economic models and the concentration of power in the hands of a few corporations. For example, transformative justice can promote the redistribution of resources and the inclusion of marginalized communities in the decision-making process related to climate change policy. As such, while particular instances of ecocide should be followed by reparations, broader transformative change is also necessary to respond to this extremely harmful crime.

Conclusion

In light of injustices faced by marginalized communities, it becomes clear that criminalizing ecocide and implementing restorative and transformative justice mechanisms could offer significant solutions. Legally prohibiting the destruction of the environment and ensuring accountability for those responsible helps protect the ecological integrity of vulnerable regions, effectively safeguarding the rights and well-being of the Indigenous community. By implementing legal measures that prohibit the destruction of the environment and holding individuals accountable for their actions, we can effectively protect the ecological integrity of vulnerable regions and ensure the rights and well-being of the Indigenous community. This approach not only addresses the specific concerns communicated in the UN Permanent Forum on Indigenous Issues, but it also aligns with the broader advocacy for holding corporate entities, states, and other powerful actors accountable for their destructive actions.

It is crucial to acknowledge that ecocide goes beyond the destruction of the natural world; it affects the air we breathe, the water we depend on, and the land upon which our existence relies. The consequences of ecocide are even more severe for marginalized communities and countries in the Global South, which often lack the resources to protect themselves from the harmful impacts of environmental degradation. By criminalizing ecocide and pursuing restorative and transformative justice, these vulnerable groups may have a viable path for seeking redress and rebuilding their lives and livelihoods. This comprehensive response recognizes the interconnectedness of climate change, crime, and violence, offering a transformative approach that aims to restore justice and confront the most pressing social problem of our time.

Notes

1 Such actions are being taken in California, where the state is suing some of the world's largest oil and gas companies. They are alleging that these companies deliberately engaged in "decades-long campaign of deception" about climate change and the dangers "posed by fossil fuels that has forced the state to spend tens of billions of dollars to address environmental-related damages" (Sahagun 2023).
2 To be certain, criminalization is likewise impacted by public attitudes and norms as well as by power dynamics. Put simply, those with power are able to define crimes and the appropriate responses to these crimes.
3 The fact that harms to property were criminalized before harming the environment and everything that lives within it is arguably due to the dominance of capitalistic worldviews.
4 See the Center for Biological Diversity, "Global Warming and Endangered Species Initiative," at www.biologicaldiversity.org/campaigns/global_warming_and_endangered_species/index.html
5 Indeed, some have suggested that we need special tribunals for environmental crimes in which judges are specially trained to respond to environmental crimes (Nurse 2016).
6 See, for instance, the International Institute for Environment and Development's work on reparations.

References

Agwu, Julia and A. A. Okhimamwe. 2009. "Gender and Climate Change in Nigeria." *Lagos, Nigeria: Heinrich Böll Stiftung (HBS)* 71.

Annecke, Wendy. 2002. "Climate Change, Energy-Related Activities and the Likely Social Impacts on Women in Africa." *International Journal of Global Environmental Issues* 2(3/4): 207–222. https://doi.org/10.1504/IJGENVI.2002.002400

Apel, Alexis B. and James W. Diller. 2017. "Prison as Punishment: A Behavior-Analytic Evaluation of Incarceration." *The Behavior Analyst* 40(1): 243–256. https://doi.org/10.1007/s40614-016-0081-6

Babakhani, Erfan. 2023. "On the Effectiveness of Restorative Justice in the Ecocide Crime." *Vilnius University Open Series* 7–15. https://doi.org/10.15388/PhDStudents Conference.2023.1

Bassadien, Shahanna Rasool and Tessa Hochfeld. 2005. "Across the Public/Private Boundary: Contextualising Domestic Violence in South Africa." *Agenda* 1(1): 4–15.

Bui, Hoan and Merry Morash. 2008. "Immigration, Masculinity, and Intimate Partner Violence from the Standpoint of Domestic Violence Service Providers and Vietnamese-Origin Women." *Feminist Criminology* 3(3): 191–215. https://doi.org/10.1177/1557085108321500

Chaffin, Brian C., Ahjond S. Garmestani, Lance H. Gunderson, Melinda Harm Benson, David G. Angeler, Craig Anthony Arnold, Barbara Cosens, Robin Kundis Craig,

J. B. Ruhl and Craig R. Allen. 2016. "Transformative Environmental Governance." *Annual Review of Environment and Resources* 41(1): 399–423. https://doi.org/10.1146/annurev-environ-110615-085817

Crook, Martin and Damien Short. 2014. "Marx, Lemkin and the Genocide – Ecocide Nexus." *The International Journal of Human Rights* 18(3): 298–319. https://doi.org/10.1080/13642987.2014.914703

Crook, Martin, Damien Short and Nigel South. 2018. "Ecocide, Genocide, Capitalism and Colonialism: Consequences for Indigenous Peoples and Global Ecosystems Environments." *Theoretical Criminology* 22(3): 298–317. https://doi.org/10.1177/1362480618787176

Department of Justice. 2015. "US and Five Gulf States Reach Historic Settlement with BP to Resolve Civil Lawsuit over Deepwater Horizon Oil Spill." Accessed June 30. www.justice.gov/opa/pr/us-and-five-gulf-states-reach-historic-settlement-bp-resolve-civil-lawsuit-over-deepwater

Dölling, Dieter, Horst Entorf, Dieter Hermann and Thomas Rupp. 2009. "Is Deterrence Effective? Results of a Meta-Analysis of Punishment." *European Journal on Criminal Policy and Research* 15(1–2): 201–224. https://doi.org/10.1007/s10610-008-9097-0

Gensburg, Lenore J., Cristian Pantea, Edward Fitzgerald, Alice Stark, Syni-An Hwang and Nancy Kim. 2009. "Mortality Among Former Love Canal Residents." *Environmental Health Perspectives* 117(2): 209–216. https://doi.org/10.1289/ehp.11350

Higgins, Polly, Damien Short and Nigel South. 2013. "Protecting the Planet: A Proposal for a Law of Ecocide." *Crime, Law and Social Change* 59(3): 251–266. https://doi.org/10.1007/s10611-013-9413-6

Kaminski, Isabella. 2023. "Growing Number of Countries Consider Making Ecocide a Crime." *The Guardian*. Retrieved October 31, 2023. www.theguardian.com/environment/2023/aug/26/growing-number-of-countries-consider-making-ecocide-crime

Lynch, Michael and Michael Long. 2022. "Green Criminology: Capitalism, Green Crime and Justice, and Environmental Destruction." *Annual Review of Criminology* 5(1): 255–276. https://doi.org/10.1146/annurev-criminol-030920-114647

Mann, Jessee and Baffour Takyi. 2009. "Autonomy, Dependence or Culture: Examining the Impact of Resources and Socio-cultural Processes on Attitudes Towards Intimate Partner Violence in Ghana, Africa." *Journal of Family Violence* 24(5): 323–335. https://doi.org/10.1007/s10896-009-9232-9

Nixon, Rob. 2011. *Slow Violence and the Environmentalism of the Poor*. Cambridge, MA: Harvard University Press.

Nurse, Angus. 2016. *An Introduction to Green Criminology and Environmental Justice*. City Road, London: SAGE Publications Ltd. https://doi.org/10.4135/9781473971899

Rahman, Md. Sadequr. 2013. "Climate Change, Disaster and Gender Vulnerability: A Study on Two Divisions of Bangladesh." *American Journal of Human Ecology* 2(2): 72–82. https://doi.org/10.11634/216796221302315

Sahagun, L. 2023. "California Sues Five Major Oil Companies for 'Decades-long Campaign of Deception' about Climate Change." *Los Angeles Times*. Accessed November 17, 2023. https://www.latimes.com/california/story/2023-09-16/california-sues-five-major-oil-companies-for-lying-about-climate-change

13

FRAMING THE CLIMATE CRISIS

A sociological lens through documentary film

Cedric A. L. Taylor

The problem: There needs to be an understanding and acknowledgment of the human dimension of the climate crisis. Purely scientific and technological approaches are by themselves unable to fully engage the public and policymakers and consequently move the needle on the issue.

The solution: Integrating sociological insights with the compelling storytelling of documentary filmmaking offers a more holistic and engaging approach. Sociology provides an in-depth analysis of societal behaviors and patterns relating to climate change, while documentaries translate these insights into accessible and emotionally resonant narratives.

Where this has worked: Documentaries like *An Inconvenient Truth* and *Chasing Ice* have successfully raised public awareness and influenced environmental policy by combining scientific information with human stories. Aspiring filmmakers preoccupied with climate change can make sociologically informed films that have the potential to generate similar impact.

The urgency for climate action has never been greater. The burning of fossil fuels and deforestation are primary causes of the planet's average temperatures rising at an alarming rate. We are experiencing the impacts of a warming world in myriad of ways including heat waves, droughts, wildfire smoke, heavy rainfall, flooding, and shifting wildlife populations and habitats. These impacts have, in turn, broad and dire implications for food and water resources, human health, economies, and social systems, particularly for the most vulnerable populations. The evidence suggests that efforts to address these climate challenges have not been enough. There

DOI: 10.4324/9781003437345-18

is a dire need for comprehensive approaches to move toward reducing greenhouse gases, adapt to the impacts of climate change, and transition to a more sustainable and resilient future. This is no small feat, given that the complexities around climate change are not only environmental or technological, but also behavioral, societal, and cultural. An interdisciplinary approach, combining sociological perspectives with the power of documentary filmmaking, is critical in creating effective messaging around the climate crisis.

Sociology, with its focus on societal patterns, beliefs, and behaviors, offers valuable insights into how communities perceive, are affected by, and respond to climate change. It highlights the social construction of climate knowledge, as well as issues of environmental inequality and justice, and explores the dynamics of risk perception and community resilience. These sociological lenses allow us a holistic understanding of the climate crisis.

On the other hand, documentary films have emerged as a powerful medium for communicating complex issues like climate change to a broad audience. Their ability to immersively tell stories can bring sociological insights to life, making the abstract real. Documentary films hold a mirror to society and have the capacity to raise awareness, foster discussions, and spur action.

This chapter outlines how integrating sociological understanding and documentary storytelling can create new pathways for addressing the climate crisis. It discusses the potential of this interdisciplinary approach to deepen public understanding, influence policy, and drive collective action toward sustainable solutions. In the following sections, I speak to some sociological contributions to climate change discourse, the power of documentary film in shaping public consciousness, and how these fields can together contribute to a necessary and effective response to one of the most pressing challenges of our time. I do this while making references to several films and sharing practical tips for the benefit of anyone looking to produce a climate change documentary in the future.

Understanding the climate crisis through a sociological lens

The climate crisis, though often framed mainly as an environmental or scientific challenge, is inextricably linked to the social. Sociological perspectives provide a vital framework for understanding this complex issue, revealing how social, cultural, and institutional realities influence both the causes and consequences of climate change.

First, it should be clear that the way societies perceive and understand climate change is not solely based on scientific facts but rather on cultural stories, media representations, and political ideologies (Shi et al. 2015). Many scholars are preoccupied with examining these social constructions, highlighting how different communities conceptualize climate change based on their cultural beliefs, values, and experiences (Pettenger 2007). This perspective is crucial in understanding the diverse responses to climate science and policy, ranging from climate activism to skepticism to outright denial.

Second, there is conclusive evidence showing that climate change does not impact all communities equally. Marginalized and socioeconomically disadvantaged groups often bear the brunt of climate impacts despite contributing least to its causes (Onbargi 2022). This sociological lens brings to light issues of climate justice, emphasizing the need for policies that address not only the environmental dimensions of climate change but also its social inequities.

Third, how individuals and communities perceive the risks associated with climate change significantly influences their behavior and policy preferences. Sociologists investigate these perceptions, exploring how factors like personal experience, trust in scientific institutions, and sociocultural background affect attitudes toward climate risks. Understanding these dynamics is important for effective communication strategies that resonate with diverse audiences and encourage proactive responses to the climate crisis.

Finally, there is sociological research that focuses on community resilience, exploring how social networks, local knowledge, and institutional support contribute to adaptive capacities. This research is critical in designing interventions that strengthen community resilience and empower local responses to climate challenges.

Overall, sociological perspectives around the climate crisis offer a holistic view that complements environmental and scientific perspectives. It stresses the importance of considering human dimensions in understanding and addressing the challenges of climate change. This of course is critical for developing comprehensive climate strategies that are equitable, culturally sensitive, and socially sustainable.

The role of documentary film in climate discourse

Documentary film, which is an effective vehicle for storytelling and persuasion, has played an important role in shaping public discourse on climate change. One of the main strengths of documentary films is the ability to transform abstract, often complex sociological and scientific concepts into compelling stories. These stories make the intangible aspects of climate change – such as greenhouse gas emissions, rising temperatures, and sea-level rise – tangible and relatable to the viewer. Statistics and scientific evidence are crucial for understanding the magnitude and urgency of climate change, but they often fail to resonate on a personal level. Data on rising temperatures, or melting ice caps, are necessary for understanding the scope and urgency of climate change. However, raw data and statistics might feel abstract and impersonal to many people, what psychologists call the "identifiable victim effect" (Lee and Feeley 2018). People tend to preferentially respond to the suffering of a single, identifiable individual as opposed to a large, anonymous group. This is reflected in the popular adage "one death is a tragedy;

a million deaths are a statistic." By juxtaposing data and statistics *with* personal stories, documentaries humanize the climate crisis, connecting global phenomena to individual experiences and emotions. This narrative approach is instrumental in breaking down cognitive barriers that often make it challenging for the public to engage with scientific data and statistics.

The award-winning documentary *An Inconvenient Truth* (2006) directed by Davis Guggenheim, is a prime example. The documentary film centers around former vice president Al Gore and his campaign to educate the public about the reality and dangers of climate change. The bulk of the film sees Gore presenting to an audience, a slide show laden with comprehensive statistics and data on the sources of carbon emissions, species extinction, the relationship between carbon dioxide (CO_2) levels and temperature, and melting polar ice caps and glaciers. At film school, I recall my instructor warning us against using data and graphs in our films. "The audience will check out," she said. However, *An Inconvenient Truth* (2006) somehow works superbly for three main reasons. First, Gore explains complex concepts in an accessible way, making it easy for anyone to understand the gravity of the situation. Second, he not only presents the scientific evidence for global warming, but he highlights the political and economic forces that contribute to the problem. Third, and perhaps most importantly, the film intersperses clips of the presentation with personal reflections on his life and political career and behind-the-scenes footage of his efforts to convey his message to various audiences.

Gore's reflections help create a sense of connection and empathy, making the audience more receptive to the message being conveyed. His stories also lend credibility to his message. They don't just see a former vice president; they see a public servant who has had a long-standing commitment to environmental issues. By sharing his own experiences and struggles in trying to raise awareness about climate change, Gore demonstrates a genuine concern and commitment to the issue, which then helps build trust with the audience. This, by the way, is a remarkable feat given Gore's prior reputation as "wooden" (Newfield 1999). Aside from gaining the audience's trust, the juxtaposition of personal stories with scientific data maintains a compelling narrative structure. This approach has the practical function of keeping the audience engaged and breaking up dense scientific information, making the content more digestible for a general audience.

Documentaries use powerful visual storytelling techniques that can clearly show the impacts of climate change. This is particularly relevant in the context of climate change, a crisis that is often abstract and overwhelming for a significant portion of the population. Using striking visuals in a documentary can make the complexities of climate change appear clearer and more real. Whether it is footage of wildlife struggling to survive, clear and concise infographics showing alarming trends, or animations that explain complicated processes, visuals play a crucial role in making the documentary impactful. Two extraordinary examples are *Chasing Ice* (2012) and *Chasing Coral* (2017), both directed by Jeff Orlowski.

Chasing Ice (2012) is a documentary that follows the efforts of environmental photographer James Balog and his team to visually document the melting glaciers in the Arctic. After setting up 25 cameras in front of several glaciers, Balog used time-lapse photography over 3 years to capture the dramatic changes occurring in the Arctic glaciers. Incidentally, time-lapse photography is a technique where a series of photos are taken at regular intervals over a period and then combined to create a video that shows a sped-up version of the events captured. It is a way to see slow processes happen quickly so that changes may be seen more easily. Because we are predisposed to believe what we see, the striking images have the effect of giving audiences tangible evidence. *Chasing Coral* (2017) highlights the beauty and importance of coral reefs, which are home to a quarter of all marine species and provide food and livelihoods for millions of people. However, the film reveals that coral reefs are dying at an alarming rate due to rising sea temperatures and other human-induced stresses. Stunning underwater cinematography and time-lapse photography show the "before and after" of coral bleaching. These films show images that have a lasting impact, that stir emotional responses and foster a deeper understanding of the urgency of the climate crisis.

Documentaries also highlight the human dimension of climate change by show-casing stories of environmental inequality, social injustice, and community resilience. A great example is the film *Cooked: Survival by Zip Code* (2019), directed and produced by Judith Helfand. It follows the 1995 Chicago heat wave that resulted in over 700 deaths in one week and primarily affected poor and disadvantaged communities. The film highlights how long-standing inequities result in poorer Black and Brown neighborhoods being vulnerable during such climate crises. Hefland's film not only exposes inequalities but uses stories of individuals to create a connection and understanding among viewers.

A personal favorite of mine, *Mossville: When Great Trees Fall* (2020), directed by Alexander Glustrom, also comes to mind. That film explores environmental racism in the United States, focusing on a centuries-old Black community (established by freed slaves) in Louisiana that is uprooted by petrochemical plants. There are many more documentary films out there that focus on how climate change disproportionately affects marginalized communities.

Documentaries don't just educate – they also inspire viewers to learn more, discuss, and act. Postscreening discussions, social media campaigns, and community events tied to documentaries create spaces for public discourse, encouraging collective reflection and action. A great example is the short film *The Mud on Their Hands* (2022), directed by Jason Whalen, which features Tyronne Edwards, a pastor in Phoenix, Louisiana, who led his community's rebuilding efforts after Hurricane Katrina and is now working to help his community protect their homes from future hurricanes. My own documentary, *Nor Any Drop to Drink: Flint's Water Crisis* (2018), was also critical in creating a more informed and engaged public, capable of driving change around water regulations at the state level. Documentary films that command significant public and media attention can potentially embarrass or pressure the powers that be to "do the right thing."

Integrating sociological insights in the filmmaking process

Occasionally, I teach courses and workshops on documentary filmmaking to young scholars and activists who are passionate about a range of environmental issues. More than producing their own technically sound film projects, my students want their own films to make an impact. There are a lot of things that go into making an impactful film, and there are a lot of ways to do it. However, as a sociologist working in the filmmaking space, I can recommend several strategies that can help filmmakers create impactful films on climate change as well as other social issues documentary.

Tip 1: Do thorough research. Yes, this is a given for any filmmaker in the preproduction phase. However, doing extensive research into the sociological aspects of climate change is necessary if you want the film's content to convey more than the scientific and technological aspects of the issue. This step means understanding how different communities are affected, societal attitudes toward climate change, and existing inequalities. In preproduction, filmmakers likely do preliminary interviews with a variety of people, including experts, to gain an understanding of the topic. I recommend having conversations with environmental scientists, experts in climate policy, *and* sociologists who do climate work to ensure your film is grounded in accurate and current knowledge.

Tip 2: Find and focus on important sociological theme(s). Based on your research, I recommend focusing on themes like environmental justice or cultural attitudes toward climate change. This will give your film a clear focus and hopefully make it more impactful. Look for stories that exemplify these themes, such as communities disproportionally affected by climate change or successful local initiatives addressing climate issues.

Tip 3: Engage with communities inclusively. There is no going around it. You should spend as much time as you can in the communities you are documenting. The strength of my film *Nor Any Drop to Drink: Flint's Water Crisis* (2018) was not based on its high production value (the film was made on a shoestring budget). The film was successful, in part, because of the time spent in Flint that allowed us to build trust and relationships. Community members felt comfortable sharing their stories and understood how their participation could raise awareness of their circumstance. Audiences as a result were privy to intimate and sometimes uncomfortable truths about what being poor and Black means with respect to exposure to environmental harms.

Research points to climate change having a differential impact on different geographies and identities (e.g., gender, race, and socioeconomic status). I therefore recommend trying to represent a diverse range of voices and perspectives, especially those who are typically marginalized in the climate change discourse.

Counterstories, a concept originating from critical race theory and often used in sociology, refers to narratives in which marginalized groups can articulate their own often overlooked experiences and perspectives, which are alternative to the dominant or mainstream stories commonly accepted in society. These can be incorporated in your film. Filmmakers may want to consider incorporating collaborative approaches in the filmmaking process, where community members are actively involved in shaping their narratives.

Tip 4: Tell compelling personal stories. This is a very important one. Find individuals whose experiences personify the broader sociological issue(s) you are exploring. This approach can be used to make abstract concepts like climate change *real*, relatable, and not merely an intellectual exercise being undertaken by out-of-touch elites. It should go without saying that filmmakers should be responsible with the personal stories they are entrusted with. Don't exploit sensitive situations for dramatic effect. These stories should be told with respect and sensitivity, especially when they are from vulnerable or marginalized communities. Further, respect the dignity and autonomy of your subjects. Obtain informed consent and give them some control over how their stories are told. The last thing you want to do is to cause damage to individuals or communities that are already going through it.

Tip 5: Use data effectively. I believe that in many Western societies, people are swayed by statistics and numbers (even though they may not really understand their meaning). If you are making a film related to climate change, you should use data to support the sociological aspects of your narrative. This could come in the form of visuals and animations as well as expert interviews in your film. I have a few caveats to this tip. If you use visuals and animation, make sure they are in harmony with the overall look, feel, and tone of your film, otherwise you will jar your audience out of the story. Also don't overdo it. Too much data and numbers on the screen can be very distracting. If you have expert voices in your film who can explain the scientific data behind climate change, your film will gain credibility that can counter misinformation and some level of skepticism. Of course, there are varying amounts of skill among experts with respect to communicating to the public. Hopefully, in your preproduction stage, you would have identified experts who are good at breaking down data into digestible pieces. What you don't want is an expert who can only convey information in technical jargon.

The point was made earlier in this chapter, but it is worth repeating here: You will want to contextualize the data within personal stories to make it more meaningful and impactful. For reference, I suggest that you watch *Unbreathable* (2020) directed by Maggie Burnette Stogner. That film examines the challenges the United States faces in ensuring healthy air for everyone, particularly in poorer communities, and looks at the Clean Air Act that regulates air emissions.

Tip 6: Highlight solutions and resilience. You have masterfully conveyed the problems around an aspect of climate change. Now what? Without an appropriate ending, audiences will feel cheated after having invested time and perhaps emotion in your film. In many introductory sociology classes, students learn about Social

Learning Theory that states that people learn by observing others. When individuals see stories of others achieving positive change, it can motivate them to emulate these behaviors (Bandura 1977). Documentary films typically follow the three-act structure where there is the beginning, the middle, and the end. The end is where you will craft the conclusion or resolution of the story. Considering what the sociological research suggests, I would recommend highlighting resilience, successful initiatives, presenting a call to action, or reflecting on the journey and the road ahead. The end should provide a sense of closure by answering the questions posed at the beginning and showing a way forward. You could choose to end on a pessimistic or grim note, but I recommend having an ending that leaves the audience feeling inspired and motivated to act. Joshua Tickell's and Rebecca Harrell Tickell's film *Kiss the Ground* (2020) argues that the current agricultural system is broken and that conventional farming inclusive of pesticide use, tilling, monocultures, and overfertilizing are problematic. Pretty grim. However, the film ends with a segment of images of large areas of land where desertification has been reversed. It ends on a hopeful note, describing solutions that are being implemented to promote regenerative agriculture that can not only halt climate change but also reverse some of its effects.

Conclusion

The integration of sociological insights with documentary filmmaking offers a powerful approach to addressing the climate crisis. Sociology, with its focus on societal patterns, beliefs, and behaviors, provides a comprehensive understanding of the human dimensions of climate change. It highlights the diverse ways communities perceive, are affected by, and respond to this threat, emphasizing the importance of considering environmental inequality, justice, risk perception, and community resilience. Documentary films, on the other hand, translate these sociological concepts into compelling narratives, making the abstract real and relatable to a broader audience. They can also be used to shape and shift the troubling cultural stories that perpetuate exploitative and extractive practices, such as those explored by Osborn, Pellow, and Haltinner in this volume. Film also influences the way stories about climate change are heard, and doing so in ways that tap into sociological knowledge of emotions and identity, as discussed by Sarathchandra in Chapter 9. Sociology offers depth in understanding, while documentary filmmaking ideally provides the accessibility and emotional impact needed to engage a wider audience. By harnessing both sociology and documentary filmmaking, a more informed and engaged public, capable of driving the social change needed to address the climate crisis, is made possible.

References

Bandura, Albert. 1977. "Self-efficacy: Toward a Unifying Theory of Behavioral Change." *Psychological Review* 84(2): 191–215. https://doi.org/10.1037/0033-295X.84.2.191

Fennell, Judith Helfand (Director). 2019. *Cooked: Survival by Zip Code*. Chicago, IL: Kartemquin Films.

Glustrom, Alexander John (Director). 2020. *Mossville: When Great Trees Fall*. United States: Watchers of the Sky LLC.

Guggenheim, Davis (Director). 2006. *An Inconvenient Truth*. United States: Paramount Classics.

Lee, Seyoung and Thomas Hugh Feeley. 2018. "The Identifiable Victim Effect: Using an Experimental-Causal-Chain Design to Test for Mediation." *Current Psychology (New Brunswick, N.J.)*. New York: Springer US.

Newfield, Jack. 1999. "Gore Goes from Wooden to Plastic." *New York Post*, October 18. Retrieved November 16, 2023 (https://nypost.com/1999/10/18/gore-goes-from-wooden-to-plastic/).

Onbargi, Alexia Faus. 2022. "The Climate Change – Inequality Nexus: Towards Environmental and Socio-Ecological Inequalities with a Focus on Human Capabilities." *Journal of Integrative Environmental Sciences*. Abingdon: Taylor & Francis.

Orlowski, Jeff (Director). 2012. *Chasing Ice*. United States: Submarine Deluxe.

Orlowski, Jeff (Director). 2017. *Chasing Coral*. United States: Exposure Labs.

Pettenger, Mary E. (Ed.). 2007. *The Social Construction of Climate Change: Power, Knowledge, Norms, Discourses*. New York: Routledge.

Shi, Jing, Vivianne H. M. Visschers and Michael Siegrist. 2015. "Public Perception of Climate Change: The Importance of Knowledge and Cultural Worldviews." *Risk Analysis* 35(12): 2183–2201.

Stogner, Maggie Burnette (Director). 2020. *Unbreathable: The Fight for Healthy Air*. United States: Center for Environmental Filmmaking, American University.

Taylor, Cedric (Director). 2018. *Nor Any Drop to Drink: Flint's Water Crisis*. United States: Central Michigan University.

Tickell, Josh and Rebecca Harrell Tickell (Directors). 2020. *Kiss the Ground*. United States: Big Picture Ranch.

Whalen, Jason (Director). 2022. *The Mud on Their Hands*. United States: Independent Film.

PART 5

Organizing through a new ethic

14

FROM VULNERABILITY TO CO-PRODUCTION

Centering Indigenous ecologies in Arctic climate adaptation

P. Joshua Griffin

The problem: Indigenous peoples, nations, and communities are among those most negatively impacted by runaway climate change, which also amplifies the structural violence, economic precarity, and political marginalization of ongoing colonialism.

The solution: Indigenous peoples understand for themselves how best to address climate impacts while also centering long-standing goals of cultural continuance, food sovereignty, collective well-being, and resurgent political self-determination.

Where this has worked: Through long-term relationships with Indigenous communities, Arctic social (and physical) scientists can embrace the co-production of knowledge and action in support of locally identified priorities for climate adaptation.

Kivalina is a 500-person Iñupiaq (Alaska Native) community located on a barrier island between the Chukchi Sea and a brackish lagoon in Northwest Alaska. Late one evening in May 2019, I was sitting by the ocean with my dear friend Replogle "Reppi" Swan Sr. – an accomplished hunter, whaling captain, and community first responder. We spent several hours just talking and watching the ocean sparkle in the near-midnight sun. We were 80 miles above the Arctic Circle, and there was open water as far as the eye could see. *It was way too early for open water*.

Since starting as a doctoral student in 2012, I've worked on a variety of projects in Kivalina, focused on a range of environmental issues. Earlier that day, Reppi had

DOI: 10.4324/9781003437345-20

been paging through a draft of my dissertation when he suddenly looked up, "Do you remember when I told you that my boys would never know what it means to hunt from the ice?"

"Yeah," I said.

"It happened even faster than I thought." I sat silently with the weight of his statement. "You should put that in there," he added.

"Ok, I will."

Sitting by the ocean that evening, we continued our earlier conversation. "I'm trying to adapt alright," Rep declared, "but there's no ice. It's still *umiaqtuq* (whaling) season, but no ice. Whales and belugas are passing right out there, we just can't get to them."

We're quiet once again until I ask, "What do you think you need to be successful?"

"We can be . . . if we work together." Rep explained how hunting safely in open water requires larger boats than from the ice. "This is new to us," he continued, "our tradition was using *umiaq* boats – paddle boats – to hunt through a small lead that opens every year. Now look, no ice. Big open ocean . . . too big for a paddle boat . . . rapidly changing weather."

As we watched the ocean sparkle, Rep shared what he has learned from communities up north and what he might try differently in the future. Each fall in Utqiaġvik, Kaktovik, and Nuiqsut, they hunt bowheads in open water with big oceangoing boats: the crews must work together, first to strike, and then to land a bowhead. In the past, Kivalina's whalers could haul their catch onto a thick ledge of ice attached to shore. Rep's crew chased two whales this past April, but even if they struck, his 18-foot Lund boat working alone might not be enough to bring it safely home through all that open water. Talking to Rep, it's clear that Kivalina's whaling captains already know what they need to be successful under increasingly difficult and uncertain conditions.

Sea-ice loss in context

Global warming has had profound effects on individual and collective life in Kivalina, but perhaps the most severe has been to undermine the reliable presence and thickness of sea ice in the Chukchi Sea (Mahoney et al. 2014). In the Lower-48, sea-ice loss is viewed primarily as a geophysical phenomenon indicating an unfolding *global* catastrophe. From the feedback loops causing the Arctic to warm much faster than the rest of the world (Dai et al. 2019) to its role in destabilizing ocean and atmospheric circulation patterns (Vihma 2014), sea-ice loss is indeed a planetary-scale crisis. But for diverse Indigenous communities across the Arctic, the impacts of sea-ice loss are also experienced socially, at far more immediate and local scales.

When I first visited Kivalina in June 2012, the elder Jerry Norton took me to the ocean in search of the last of the season's "good coffee water." After pulling the boat up on a slab of floating sea ice, Uncle Jerry moved slowly over mounds of crystalized snow, lingering at each puddle formed from rain or melted snow on top of ice. Careful not to disturb the slush below, he tasted each pool from the dipper he held in his right hand. Upon finding one with the right clarity and sweetness, Jerry took a knee and filled the blue plastic can he held in his left hand.

Earlier that same day, Jerry and I stood on the gravel beach, looking out at the ocean. For days, the ice had been jammed up close to shore. All of Kivalina was waiting for an east wind to blow the ice away and out over the horizon. When the ice came back, Jerry told me, it would be loaded with *ugruk* (bearded seal). Sure enough, as the ice dispersed and then returned, hunters would fill their family's drying rack with enough black meat to last the year. In 2019, however, Kivalina's hunters traveled over 50 miles northwest to find few *ugruk* among the broken ice just past Cape Thompson.

Throughout the Arctic, the pace of sea-ice loss is faster than scientific models predicted even a few years ago (Jansen et al. 2020). This loss of habitat creates new challenges for ice-dependent species and Iñupiaq hunters alike (Hauser et al. 2021). While sea-ice loss follows an overall downward trend, sea-ice extent and thickness are not linear from year to year. Kivalina's ocean hunters remain ready for the better years, like June 2022 and 2023 when broken chunks of ice offered migrating *ugruks* ample refuge just offshore.

While climate change has created new challenges for *ugruk* hunting, the loss of sea ice has been most disruptive to Kivalina's bowhead whaling practices. Until the early 1990s, whaling crews would camp for weeks on the ice some 10 to 20 miles offshore. This frozen platform positioned them along a migration path where whales would surface to breathe in small openings in the ice, called leads. Today, Kivalina's whaling captains are lucky to camp a few nights just a mile out from town. With all the open water now, bowheads can surface almost anywhere to breathe – thick black needles in a dark and watery haystack. Back in the day, the ice was no guarantee for whaling success, but it provided context for a good faith return on one's patient preparations and labor (Griffin 2020b). Bowhead whaling is now more challenging, but as Reppi can attest, as long as the whales pass by, the practice will endure.

From vulnerability to co-production

In the last two decades, popular and scholarly representations of climate change have cast the Arctic as an icon of unfolding global loss. Within this discourse, Arctic Indigenous communities and cultures have been portrayed as especially vulnerable to rapid environmental change (Callison 2017). Even the most sensitive journalistic accounts of Kivalina's ingenuity and resilience have been subjected to editorial framings of disappearance and vanishment (Knafo 2019). Much of the earliest scholarship on Arctic climate adaptation has been similarly deterministic, erasing potential insights into Indigenous creativity through the lens of "vulnerability" (Ford et al. 2008). After saturating the literature, vulnerability frameworks were justifiably critiqued as disempowering and antipolitical (Cameron 2012). Prominent vulnerability scholars have since come to recognize the importance of history, political economy, and colonialism (Ford et al. 2016).

Despite the early dominance of vulnerability approaches, alternate frameworks to Arctic climate adaptation have persisted. A resilience standpoint, for example, might begin with a community's adaptive strengths and ask what resources or networks could enhance "the capacity for learning and self-organization" across multiple scales (Berkes and Jolly 2001). Values-based approaches also center local and community perspectives as to which "adaptation pathways are perceived as most desirable, effective, and legitimate" (O'Brien and Wolf 2010). Importantly, resilience and values-based approaches resonate with the methodologies of critical Indigenous studies, such as Eve Tuck's long-standing call "to craft our research to capture *desire* instead of damage" (2009: 416).

The last decade has brought wide and welcome changes in how many professional Arctic researchers engage with Indigenous peoples. This is especially evident in how physical and social scientists are working in collaborative ways with Inuit and Iñupiaq communities to better understand the dynamics of sea ice, ice-associated species, and the concerns of ice-connected peoples (Hauser et al. 2023; Wilson et al. 2021; Fox et al. 2020). The earliest efforts in this regard were focused on including Indigenous peoples' expert knowledges (i.e., traditional ecological knowledge, or "TEK") as a complement to the observations and predictions of professional scientists. While expanded recognition of "TEK" was promising, many such endeavors remain limited by a compartmentalized treatment of Indigenous knowledge – that is, the splitting off of ecological or environmental "data" from the associated norms, values, and obligations of holistic knowledge systems (Nadasdy 1999).

In contrast, some scholars in these spaces began not only to ask what sea-ice change might *mean* to Arctic Indigenous communities (Gearheard et al. 2013) but to also design research with community interests in mind (Gearheard et al. 2011; Laidler et al. 2011; Druckenmiller et al. 2013). The proliferation of community-based monitoring over the last decade has begun to mainstream more equitable knowledge production between polar scientists and Arctic Indigenous communities (Johnson et al. 2016). Many such initiatives not only track Arctic change in a more holistic way but also support "community priorities and local-scale needs" (Hauser et al. 2023). Efforts to co-develop the questions, methods, and *goals* behind sea-ice research are today described as the "co-production of knowledge" (Druckenmiller 2022; Wilson et al. 2021; Fox et al. 2020).

Kivalina Volunteer Search and Rescue

A dark ramp of sand flows west from Kivalina's rock revetment wall down to the edge of the Chukchi Sea. On days like this, the water laps gently, but far from the water, scattered clusters of round smooth stones testify to a much angrier ocean.

The rock revetment was constructed on Kivalina's western shore by the US Army Corps of Engineers 15 years ago to provide partial protection from dangerous erosion during fall storms – all exacerbated by the late-forming sea ice (Fang et al. 2018). How could the source of such rich abundance also yield such tumult?

Two boats are held to shore by a line attached to an anchor buried some 20 feet up the bank. The hull of Reppi's 18-foot aluminum Lund was originally red but has been spray-painted white as camouflage from sea mammals. Rep crouches in the stern, floating just off the bottom while the bow rests securely in the sand. He is jostled lightly as he tinkers with a hose connection to the outboard motor. It's misty and overcast, but not cold: nonetheless we're all dressed for winter.

Three hundred yards south, Mikey's Carolina Skiff emerges from the mouth of the channel and slowly motors toward us before pulling up alongside David. The fleet is now complete, and the rest of our group assembles: a dozen volunteers from Kivalina Volunteer Search and Rescue (KVL-SAR), myself, three of my graduate students, and Dorothy Adler – a well-known wilderness medical educator from Palmer, Alaska. Seventeen of us pile into three boats. Each is just a bit overloaded, but it's a short journey and good weather. After 3 years of dreaming and hard work, it's a great feeling to be together and headed out to the ice.

KVL-SAR is a group of volunteers organized to assist and aid distressed, lost, or injured hunters and travelers, whether on land or sea. The organization plays a critical role in supporting Kivalina's hunting and fishing practices, which are central for collective food sovereignty, intergenerational education, and intracommunity care (Griffin 2020a). KVL-SAR volunteers also play a vital public safety role for the City of Kivalina, as first responders to emergencies in town: from traumatic injuries to storm-induced coastal erosion (Griffin 2023).

Joe Swan Sr. traces the origins of KVL-SAR to the late 1960s when he purchased one of the first snow machines, or "snogos," in Kivalina. Before snogos, when a hunter had lost his way in a whiteout or began to suffer fatigue, he could simply lay down in the sled and let his dog team find their way home. With snogos, day trips became the norm. Hunters could now travel farther in a single day but could also break down, run out of gas, and get really lost in bad conditions. Nowadays, Joe tells me, if a hunter is out past midnight, "they're overdue."

Under his father's leadership, Reppi began volunteering with KVL-SAR in the 1990s when he was in high school. When Joe saw that Rep had learned what he needed to know – all about the land and weather – he turned the organization over to his youngest son. "When we go out searching," Rep explains, "we have to know the land by heart." He continues:

If it's stormy you remember where you are based on those landmarks – whether a willow or a rock or a creek. Even searchers get lost when it gets bad out . . . but if you keep an eye on where those landmarks are you'll know where you're headed.

The knowledge of Kivalina's territory includes many geographies: from tundra to rivers and the ocean. Land-based hazards have always varied by season, but today spring is the most unpredictable. "When you try to cross the river in springtime," Rep explains "you can't tell if its snow or slush. Most of the time it's too late, we're stuck right there."

As a whaling captain, Rep has a wealth of stories from traveling and camping on the sea ice.

> Twenty-five years ago our ice would be twenty to thirty feet thick . . . ten years ago it was half that . . . and now, today's sea ice . . . the thickest I've seen is four feet, but mostly three. So it's a real big change on how we do our hunting in the ocean . . . I could look at the ice – if it's white it's still good – but then I could just fall through.

All of these changes have led KVL-SAR to pursue new forms of preparedness and strategic growth. As KVL-SAR president, Reppi works closely with Colleen Swan, a seasoned administrator with more than three decades experience in tribal and municipal government, planning, and emergency incident command. Colleen and Reppi are working to build a more resilient and proactive organization, equipped with the knowledge, skills, and resources to anticipate and prevent dangerous incidents before they occur. Their goals range from identifying and accessing new equipment, funding streams, and office and storage space, to creating new knowledge networks for skill-based training and emerging knowledge tools to assess and anticipate environmental change.

Polar Science at a Human Scale

In September 2019, I joined the University of Washington faculty in the Department of American Indian Studies and the School of Marine and Environmental Affairs. As a graduate student, I had dreamed of one day facilitating conversations between Arctic scientists and friends in Kivalina, but as faculty within the College of the Environment, I now had colleagues working on the science of Arctic environmental and ecological change. Cecilia "CC" Bitz is a leading sea-ice forecaster and at the time was chair of Atmospheric Sciences, so I was thrilled and grateful when she agreed to meet. I told CC about my conversations with Reppi and other whaling captains over the years, along with the changes I had myself observed in Kivalina's cryosphere since 2012. "Could she create something of local relevance to community decision-making around sea-ice use?" CC explained that she and others in the forecasting community were wondering if their products could play a role in supporting the information needs of local communities. How might sea-ice users, such as hunters in Kivalina, more directly benefit from the kinds of science she was doing? What could the science offer, and what were its limits? We had independently arrived at the same question, just from different starting places.

In December 2019, I returned to Kivalina and shared with Rep and Colleen what I had learned from CC. Together we outlined the goals, methods, and budget proposal for "Polar Science at a Human Scale" (PSHS) – a co-directed collaboration between KVL-SAR, the City of Kivalina, and the University of Washington polar science community.[1] Since June 2020, our team of co-directors, multidisciplinary investigators, graduate students, and community experts has focused on co-producing both knowledge and action to address the concerns and capacity-building priorities of KVL-SAR.

To understand the state of environmental research at Kivalina, our team divided up literature review topics like oceanography, hydrology, permafrost, along with marine mammal, caribou, fish, and bird ecology. In spring 2021, we organized a ten-week seminar with presentations from a rotating cadre of expert scientists. This helped our team build a shared understanding of the Kivalina's environmental and ecological dynamics. Most importantly perhaps, conversations between our invited specialists, Colleen, and Reppi, helped to identify the limits of existing research and to generate new research questions with relevance to KVL-SAR and the community as a whole. By winter 2022, our team had drafted a strategic plan outlining how SAR might anticipate and respond to climate-induced change across the tundra, rivers, and ocean. The plan, reflective of our project's overall design, sought to cojoin material capacity building with the ongoing co-production of actionable climate research.

While undertaking this broad scoping and planning work, PSHS also simultaneously focused specific attention on sea-ice dynamics through weekly meetings to exchange observations of ice conditions and forecasts throughout the 2020–2021 season. This process helped to identify the strengths and limitations of existing sea-ice knowledge products – from satellite photos to the forecasts and ice maps produced by academic or government institutions. Over the next two seasons, we expanded our knowledge exchange with the "Kivalina Sea Ice Project" Facebook page, designed to share knowledge products more widely, recruit additional local observations, and solicit greater community feedback.[2]

Over the last 3 years, our PSHS team has pursued additional activities as well, such as deploying satellite tracking and communication devices to facilitate coordination of KVL-SAR incidents, hosting a three-day Science and Culture workshop in Kivalina (June 2022), and working with whaling captains to quantify shifting metrics of bowhead whaling access and safety. Finally, we are producing a series of digital short stories, the first of which was leveraged to crowdsource over $7,000 to cover the costs of Kivalina's first ever Wilderness First Responder and Marine Rescue Training in June 2023.[3]

***

As Rep puts the boat up on the ice, a volunteer springs from the bow to set the anchor in a solid spot. Monchie follows suit, leaping from Mike's boat with a

5-foot long seal hook in hand. He walks forward in concentric semicircles from our landing spot, probing the ice every few steps. He drives the handle into each puddle he encounters. It sinks 8 inches here, 12 inches there, but usually hits some solid ice below. As his perimeter reaches about 30 feet, we hear him shout "Whoaaaa." Three feet of the handle just disappears like that. Monchie leans over and reaches a little farther, it drops another foot. We all gasp with nervous laughter and someone shouts, "Careful bro!" Monchie returns to us shaking his head, and we all know now not to wander.

Rep walks carefully southward along the ice edge, carefully poking and probing with his pickax, before kneeling down to chip away at a hidden overhang. A captain's first duty is the safety of his crew. By the time Dorothy and I are out of the boat, someone is already carrying her bag – filled with ropes and hardware typical of glacier travel but which she's adapted to ice and water rescue in marine environments. Yesterday, we practiced her pulley system on land, and it was strong enough to drag a parked four-wheeler through the gravel. I marvel at the scene: 17 souls floating on an iceberg in mid-June. I ask Colleen how long it has been since she was out here. Maybe 10 years or more? I have some video clips she took in 2011, while bouncing in a sled behind Dolly's snogo through rough but soggy ice. It was the last time she felt safe going out to whaling camp.

Dorothy kneels down to build an anchor of ice screws as Monchie and Rep lean over to observe. David brings over a spare Honda tire, dresses it in a lifejacket, ties it off, and casts it out to sea. We use pulleys to drag the tire back up and over the ice edge. Monchie ties himself a harness and prussock with cordalette then walks the line until he reaches the tire. "I'm here bud," he reassures our rubber friend, "You'll be ok." We all chuckle. We begin to haul the tire, but it offers no resistance. Flipping the script, Monchie belly flops on top of it and starts calling, "Help me guys! Help me!" *Now* we're simulating a rescue.

The volunteers take turns adding loops to the system and it becomes easier to drag him across the ice. We cheer and celebrate when Monchie (and the tire) are safely back to us. Now Dorothy wants to demonstrate how strong the system is, so we set a new anchor, fill a boat with people and drag them some 10 feet through the slushy ice. Before heading home, we practice the "hypo wrap" again – it's what we'd do if someone had just come out of the water and was hypothermic. Dorothy guides us through the lacing, then lays down a tarp and camping pad. "It's growing very dim," our volunteer patient groans, while his eyes flicker in and out of consciousness. "Guys," he mutters, "I'm fading out." The melodrama has us in stiches, but finding one last moment of sincerity, the team lifts him over the gunnel of Rep's boat then slides him into the bow. It's getting late. After loading up, we slowly boat around slowly to see a bit more ice. As much as everyone wants to look for *ugruks*, we're too overloaded (and cold) to go very far.

Back at the community center, we hold a debrief. The rope and pullies are powerful, but there are questions about how long it takes to set up the anchor. Would an ice axe be faster than the screws? What about an axe placement reinforced by screws? It was good to practice the hypo wrap out there – folks are really getting the hang of it now. The conversation turns to stories of past incidents and the potential scenarios we should simulate next. "You have no idea," Rep says from experience, "how hard it is to pull someone into the boat when their parkee is soaked and waterlogged. My son is *big* and he couldn't get that one guy out while I tried to keep the boat straight. It took two of us, we kept losing him." He's haunted by a memory from last spring. In the end it turned out okay, but Rep is determined never to have another close call.

We all agree how great it was to be out on the ice, but next time we should really practice with a person. "Yeah, we'll throw Griff overboard," Billy says to uproarious laughter. It's not the first time this proposal has been aired. By this point it's an idea to which we're all committed, myself included. We'll have to bring Dorothy back in a few months, when the weather gets bad.

Indigenous ecologies and climate adaptation praxis

In this chapter, I describe an ongoing collaboration with KVL-SAR to co-produce knowledge and action in support of local priorities for preparedness and resilience to climate impacts. I reflect at length on the experience of a 3-day Wilderness First Aid and marine rescue training that my students and I organized with our community co-directors and research partners. The training was an important step in expanding first responder capacity not only for environmental hazards but also the kinds of medical emergencies common in rural Alaska where basic public health and safety infrastructures are routinely deprioritized and under-resourced by federal, state, and regional agencies. Not only did the training result in eight SAR volunteers gaining Wilderness First Aid and CPR certifications, but it supported team building and cohesion through a set of shared experiences and knowledge-practices. We also built new relationships with outside partners, like Dorothy, who continues to bring new concepts, networks, and resources to our ongoing work.

Rather than "giving back" as an auxiliary aspect of "the research process," PSHS seeks to co-produce both knowledge and action by embedding community benefits and material resources within the research process itself. In this way, we approach co-production as a form of *praxis* where knowledge is generated through concrete, self-reflexive, and often experimental activity within the world. This applies equally to the technoscience we use to anticipate locally important sea-ice trends and the use of participant observation to understand *and respond to* the new challenges facing hunters, travelers, and first responders in Kivalina. Such engaged approaches to empirical research have histories within critical anthropology and Indigenous studies, where scholars like Kim TallBear call researchers to "stand with" communities through the pursuit of "share[d] goals and desires while [also] staying engaged in critical conversation and producing new knowledge and insights" (2014).

Around the world, climate change has amplified the political marginalization and structural violence that Indigenous communities face from ongoing colonialism and extractive capitalism (Curley and Lister 2020; Griffin 2020a; Whyte 2018a). This is especially evident in cases of climate-induced relocation, where colonial settlement policy is at the root of contemporary exposure to climate hazards (Griffin 2023; Maldonado 2019; Marino 2012) and settler-dominated institutions actively limit the transformational potential of community planning and infrastructure to support self-determined Indigenous futures (Jessee 2022; Shearer 2012; Marino 2009).

Kyle Whyte is a leading environmental philosopher and practitioner who brings a critical lens to the interlocking political, social, and environmental challenges facing Indigenous peoples today. Throughout his work, Whyte develops the concept of "Indigenous ecologies" to describe those dynamic constellations of social and ecological relationships that uphold Indigenous peoples' collective lives and self-governance of territory (Whyte 2018a, 2018b; Whyte et al. 2018). Indigenous ecologies, he writes (with Caldwell and Schaefer), are

> systematic arrangements of humans, nonhuman beings (animals, plants, etc.) and entities (spiritual, inanimate, etc.), and landscapes (climate regions, boreal zones, etc.) that are conceptualized and operate purposefully to facilitate a[n Indigenous] society's capacity to survive and flourish in a particular landscape and watershed.
>
> *(Whyte et al. 2018)*

For Whyte, the relationships that constitute Indigenous ecologies are normative and laden with values that guide Indigenous responsibilities, ethics, and ways of life. Importantly, Indigenous ecologies are dynamic, with the historical governance and settlement structures of many Indigenous societies following a "seasonal round" (Whyte et al. 2018) – Northwest Alaskan Iñupiat included (Burch 1998). While Indigenous peoples may themselves choose to change the ways in which they live over time and space, colonialism (especially settler colonialism) can be understood as the incursion or imposition of settler ecologies (i.e., norms of social-ecological relation) in the attempt to replace Indigenous peoples and their worlds (Whyte 2018b; 2018a). In much the same way, climate change – often in concert with colonialism – exerts a disruptive force on many critical intergenerational and self-determined Indigenous ecologies.

Within this milieu of disturbance – described by many as the Anthropocene – critical Indigenous studies scholars have highlighted how Indigenous peoples refuse, resist, or otherwise exert resilience in the face of such violent and oppressive disruptions to their ecologies. Whyte describes Indigenous efforts to not only survive but to flourish as a process of "collective continuance" (2018a; 2018b). Within the domain of climate change, and specifically climate adaptation, Indigenous communities and nations are well aware of their intersecting desires, priorities, and values. Today, diverse Indigenous peoples seek to sustain and strengthen their most valued relationships amid both climate-related and pre-climate disturbances. Through long-term relationships with Indigenous communities (and respectful partnerships with Indigenous researchers), social scientists and other allied professionals can co-produce both knowledge and action supportive of the Indigenous ecologies which uphold and further collective continuance.

Notes

1 PSHS was launched with an innovation grant from University of Washington EarthLab and has also received generous support from the University of Washington Center for American Indian and Indigenous Studies and the University of Washington Program on Climate Change.
2 See the Kivalina Sea Ice Project on Facebook: www.facebook.com/kvlseaice/
3 See: Puentes, C. "Kivalina Search & Rescue – First Responder Training." May 3, 2023, www.youtube.com/watch?v=9FyZYu8fmvQ

References

Berkes, Fikret and Dynna Jolly. 2001. "Adapting to Climate Change: Social-Ecological Resilience in a Canadian Western Arctic Community." *Conservation Ecology* 5(2): 18.
Burch Jr., Ernest. 1998. *The Iñupiaq Eskimo Nations of Northwest Alaska*. Fairbanks: University of Alaska Press.
Callison, Candis. 2017. "Climate Change Communication and Indigenous Publics." In *Oxford Research Encyclopedia of Climate Science*, edited by H. von Storch. Oxford University Press. https://doi.org/10.1093/acrefore/9780190228620.013.411
Cameron, Emilie. 2012. "Securing Indigenous Politics: A Critique of the Vulnerability and Adaptation Approach to the Human Dimensions of Climate Change in the Canadian Arctic." *Global Environmental Change* 22(1): 103–114.
Curley Andrew and Majerle Lister. 2020. "Already Existing Dystopias: Tribal Sovereignty, Extraction, and Decolonizing the Anthropocene." In *Handbook on the Changing Geographies of the State*, edited by Sami Moisio, Natalie Koch, Andrew Jonas and Christopher Lizotte, 251–262. Cheltenham, UK: Edward Elgar.
Dai, Aiguo, Dehai Luo, Mirong Song and Juping Liu. 2019. "Arctic Amplification Is Caused by Sea-Ice Loss Under Increasing CO_2." *Nature Communication* 10: 121 https://doi.org/10.1038/s41467-018-07954-9
Druckenmiller, Matthew. 2022. "Co-Production of Knowledge in Arctic Research: Reconsidering and Reorienting Amidst the Navigating the New Arctic Initiative." *Oceanography*, 189–191. https://doi.org/10.5670/oceanog.2022.134
Druckenmiller, Matthew, Hajo Eicken, Craig George and Lewis Brower. 2013. "Trails to the Whale: Reflections of Change and Choice on an Iñupiat Icescape at Barrow, Alaska." *Polar Geography* 36(1–2): 5–29. https://doi.org/10.1080/1088937X.2012.724459
Fang, Zhanpie, Patrick T. Freeman, Christopher B. Field, and Katharine J. Mach. 2018. "Reduced Sea Ice Protection Period Increases Storm Exposure in Kivalina, Alaska." *Arctic Science* 4: 525–537.
Ford, James, Barry Smit, Johanna Wandel, Mishak Allurut, Kik Shappa, Harry Ittusarjuat, Kevin Qrunnut. 2008. "Climate Change in the Arctic: Current and Future Vulnerability in Two Inuit Communities in Canada." *The Geographical Journal* 174(1): 45–62.
Ford, James, Ellie Stephenson, Ashlee Cunsolo Willox, Vcitoria Edge, Khosrow Farahbakhsh, Christopher Furgal, Sherilee Harper, Susan Chatwood, Ian Mauro, Tristan Pearce, Stephanie Austin, Anna Bunce, Alejandra Bussalleu, Jahir Diaz, Kaitlyn Finner, Allan Gordon,

Catherine Huet, Knut Kitching, Marie-Pierre Lardeau, Graham McDowell, Ellen McDonald, Lesya Nakonesznt, Mya Sherman. 2016. "Community-Based Adaptation Research in the Canadian Arctic." *WIREs Climate Change* 7: 175–191. https://doi.org/10.1002/wcc.376

Fox, Shari, Esa Qillaq, Ilkoo Angutikjuak, Dennis Tigullaraq, Robert Kautuk, Henry Huntington, Glen Liston and Kelly Elder. 2020. "Connecting Understandings of Weather and Climate: Steps Towards Co-production of Knowledge and Collaborative Environmental Management in Inuit Nunangat." *Arctic Science* 6(3): 267–278. https://doi.org/10.1139/as-2019-0010

Gearheard, Shari, Claudio Aporta, Gary Aipellee and Kyle O'Keefe. 2011. "The Igliniit Project: Inuit Hunters Document Life on the Trail to Map and Monitor Arctic Change." *The Canadian Geographer/Le Géographe Canadien* 55: 42–55. https://doi.org/10.1111/j.1541-0064.2010.00344.x

Gearheard, Shari, Lene Holm, Henry Huntington, Joe Leavitt, Andrew Mahoney, Margaret Opie, Toku Oshima, Joelie Sanguya. 2013. *The Meaning of Ice: People and Sea Ice in Three Arctic Communities*. Hanover, New Hampshire: International Polar Institute Press.

Griffin, P. Joshua. 2020a. "Pacing Climate Precarity: Food, Care and Sovereignty in Iñupiaq Alaska." *Medical Anthropology* 39(4): 333–347. 10.1080/01459740.2019.1643854

Griffin, P. Joshua. 2020b. "Thresholds" In *Anthropocene Unseen: A Lexicon*, edited by Cymene Howe and Anand Pandian. Goleta, CA: Punctum Books.

Griffin, P. Joshua. 2023. "Arctic Sea Ice Loss and Fierce Storms Leave Kivalina's Volunteer Search and Rescue Fighting to Protect Their Island from Climate Disasters." *The Conversation*. http://theconversation.com/arctic-sea-ice-loss-and-fierce-storms-leave-kivalinas-volunteer-search-and-rescue-fighting-to-protect-their-island-from-climate-disasters-191315

Hauser, Donna, Roberta Glenn, Elizabeth Lindley, Kimberly Pikok, Krista Heeringa, Joshua Jones, Billy Adams, Joe Leaveitt, Guy Omnik, Robert Schaeffer, Carla SimsKayotuk, Elena Sparrow, Alexandra Ravelo, Olivia Lee, Hajo Eicken. 2023. "Nunaaqqit Savaqatigivlugich – Working with Communities: Evolving Collaborations Around an Alaska Arctic Observatory and Knowledge Hub." *Arctic Science* 9(3): 635–656. https://doi.org/10.1139/as-2022-0044

Hauser, Donna, Alex Whiting, Andrew Mahioney, John Goodwin, Cyrus Harris, Robert Schaeffer, Roswell Shaeffer Sr., Nathan Laxague, Ajit Subramaniam, Carson Witte, Sarah Betcher, Jessica Lindsay, Christopher Zappa. 2021. "Co-production of Knowledge Reveals Loss of Indigenous Hunting Opportunities in the Face of Accelerating Arctic Climate Change." *Environmental Research Letters* 16. http://doi.org/10.1088/1748-9326/ac1a36

Jansen, Eystein., Jens Christensen, Trond Dokken, Kerim Nisancioglu, Bo Vinther, Emilie Capron, Chuncheng Guo, Mari Jensen, Peter Langen, Rasmus Pedersen, Shuting Yang, Mats Bentsen, Helle Kjær, Henrik Sadatzi, Evangeline Sessford and Martin Stendel. 2020. "Past Perspectives on the Present Era of Abrupt Arctic Climate Change." *Nature Climate Change* 10: 714–721. https://doi.org/10.1038/s41558-020-0860-7

Jessee, Nathan. 2022. "Reshaping Louisiana's Coastal Frontier: Managed Retreat as Colonial Decontextualization." *Journal of Political Ecology* 29(1): 277–301.

Johnson, Noor, Carolina Behe, Finn Danielsen, Eva Krümmel, Scot Nickels and Peter Pulsifer. 2016. *Community-Based Monitoring and Indigenous Knowledge in a Changing Arctic: A Review for the Sustaining Arctic Observing Networks*. Final Report to Sustaining Arctic Observing Networks. March 2016. Ottawa, ON: Inuit Circumpolar Council.

Knafo, Saki. 2019. "The Last Whale Hunt for a Vanishing Alaskan Village." *Men's Journal*. August 29, 2019. Accessed December 12, 2023. www.mensjournal.com/adventure/the-last-whale-hunt-for-a-vanishing-village-kivalina-alaska-w443825

Laidler, Gita, Tom Hirose, Mark Kapfer, Theo Ikummaq, Eric Joamie and Pootoogoo Elee. 2011. "Evaluating the Floe Edge Service: How Well can SAR Imagery Address Inuit Community Concerns Around Sea Ice Change and Travel Safety?" *The Canadian Geographer* 55: 91–107. https://doi.org/10.1111/j.1541-0064.2010.00347.x

Mahoney, Andrew, Hajo Eicken, Allison Gaylord and Rudiger Gens. 2014. "Landfast Sea Ice Extent in the Chukchi and Beaufort Seas: The Annual Cycle and Decadal Variability." *Cold Regions Science and Technology* 103: 41–56.

Maldonado, Julie. 2019. *Seeking Justice in an Energy Sacrifice Zone: Standing on Vanishing Land in Coastal Louisiana*. Routledge: New York.

Marino, Elizabeth. 2009. "Immanent Threats, Impossible Moves, and Unlikely Prestige: Understanding the Struggle for Local Control as a Means Towards Sustainability." In *Linking Environmental Change, Migration & Social Vulnerability*, edited by Anthony Oliver-Smith and Xiaomeng Shen. Bonn: United Nations University Institute for Environment and Human Security.

Marino, Elizabeth. 2012. "The Long History of Environmental Migration: Assessing Vulnerability Construction and Obstacles to Successful Relocation in Shishmaref, Alaska." *Global Environmental Change* 22(2): 374–381.

Nadasdy, Paul. 1999. "The Politics of TEK: Power and the 'Integration' of Knowledge." *Arctic Anthropology* 36(1–2): 1–18.

O'Brien, Karen and Wolf, Johanna. 2010. "A Values-Based Approach to Vulnerability and Adaptation to Climate Change." *Wiley Interdisciplinary Reviews: Climate Change* 1(2): 232–242.

Shearer, Christina. 2012. "The Political Ecology of Climate Adaptation Assistance: Alaska Natives, Displacement, and Relocation." *Journal of Political Ecology* 19: 174–183.

TallBear, Kim. 2014. "Standing with and Speaking as Faith: A Feminist-Indigenous Approach to Inquiry." *Journal of Research Practice* 10(2), Article N17.

Tuck, Eve. 2009. "Suspending Damage: A Letter to Communities." *Harvard Educational Review* 79(3).

Vihma, Tima. 2014. "Effects of Arctic Sea Ice Decline on Weather and Climate: A Review." *Surveys in Geophysics* 35: 1175–1214. https://doi.org/10.1007/s10712-014-9284-0

Whyte, Kyle. 2018a. "Food Sovereignty, Justice, and Indigenous Peoples: An Essay on Settler Colonialism and Collective Continuance." In *The Oxford handbook of food ethics*, edited by Anne Barnhill, Tyler Doggett and Mark Budolfson, 345–366. Oxford: Oxford University Press.

Whyte, Kyle. 2018b. "Settler Colonialism, Ecology, and Environmental Injustice." *Environment and Society* 9: 125–144.

Whyte, Kyle, Chris Caldwell, Marie Schaefer. 2018. "Indigenous Lessons About Sustainability Are Not Just for 'All Humanity'." In *Sustainability: Approaches to Environmental Justice and Social Power*, edited by Julie Sze, 149–179. New York: New York University Press.

Wilson, Katherine, Andrew Arreak, Jamesie Itulu, Gita Ljubicic, Trevor Bell and Nicole Giguère. 2021. "'When We're on the Ice, All We Have Is Our Inuit Qaujimajatuqangit': Mobilizing Inuit Knowledge as a Sea Ice Safety Adaptation Strategy in Mittimatalik, Nunavut." *Arctic* 74(4): 525–549. www.jstor.org/stable/27110558

15

CLOSING THE SOCIAL GAP IN THE DEPLOYMENT OF RENEWABLE ENERGY TECHNOLOGIES

David Bidwell and Shannon Howley

The problem: A transition from fossil fuels to renewable energy is essential to mitigate the speed and magnitude of climate change, but even with the increasing capacity of renewable energy in the United States, further implementation of renewable energy systems is necessary to lower greenhouse gas emissions. The barriers to renewable energy deployment are social as well as technological, based in part on project location and community values. While there is broad support for renewable energy, there remains localized opposition to the development of renewables projects.

The solution: Closing the gap between the broad public support of renewables and localized opposition to the siting of renewable energy technology requires involving local communities in the process while recognizing and attending to the community's values. It also requires considering the fairness and justice of the project's decision making and design. To achieve these goals, evidence-based public engagement strategies should be embraced by government agencies and energy developers.

Where this has worked: The case of the Block Island Wind Farm – the United States' first offshore wind project – exemplifies how procedural fairness and effective public engagement strategies facilitate the construction of renewable energy projects. Social science research on the Block Island Wind Farm highlights the importance of involving the local community in the project, especially trusted community liaisons and informal communication with nearby residents.

DOI: 10.4324/9781003437345-21

The mitigation of climate change requires a reduction in the burning of fossil fuels (such as coal and natural gas), which have been widely used to generate electricity. There are three pathways to reducing emissions from electricity generation:

1) Changing behaviors to lower demand
2) Increasing the efficiency of electric technologies
3) Switching to less carbon-intensive sources of generating electricity

While the first two pathways are useful, especially when many small efforts are combined, our appetite for behavior change is limited, and technological efficiency alone is insufficient for meaningful change. Most mitigation policies focus on the third pathway: the adoption of modern renewable energy technologies such as wind turbines, solar photovoltaics, and geothermal systems.

Yet, adding renewable energy to our electrical system presents social challenges. The adoption of offshore wind energy technology provides one example. In 2001, a private developer proposed to install more than 130 wind turbines in waters between Cape Cod and Martha's Vineyard, Massachusetts. This proposed development, known as Cape Wind, would have been the first offshore wind farm in the United States. What followed was more than 15 years of protests and litigation, with the developer finally abandoning the project in 2017. The failed Cape Wind project has become a symbol of public opposition to renewable energy projects. The death of Cape Wind, however, came shortly after the successful completion of a much smaller project – five turbines – off the coast of Block Island, a small and scenic New England community, in 2016.

Disputes over the development of renewable infrastructure have been common around the world. Yet, the global production of renewable electricity has been increasing steadily, albeit slowly, over the past two decades. A community of researchers has grown around the study of how the public responds to renewable energy, known as *social acceptance*. As social acceptance scholars, we research the factors that lead to support and opposition to projects like the Block Island Wind Farm, as well as what these cases tell us about potential pathways to a successful transition of energy systems.

This chapter summarizes some of the ways the public responds to renewable energy and what social science has revealed about potential pathways toward a broader adoption of renewable sources of electricity. It ends with some of the lessons we learned from studying the Block Island Wind Farm that can be employed in other communities.

Climate change and energy

Fossil fuel combustion is responsible for about 77% of US greenhouse gas emissions (Martinich et al. 2018), mainly caused by our reliance on fossil fuels for energy. Within the overall energy system, the electricity sector is responsible for about one-fourth of US total emissions. Here, about 60% of the energy used for electricity generation is provided by fossil fuels (EPA 2023a), 22% from renewable energy sources, and 18% from nuclear energy (Energy Information Administration [EIA] 2023a). With the increasing electrification of the transportation (e.g., electric cars) and other sectors, there will be greater electricity demand (EPA 2023b).

This is also true across the globe. In 2021, the global electricity sector represented over one-third of energy-related carbon dioxide (CO_2) emissions, emitting 13 gigatons of CO_2 (International Energy Agency [IEA] 2023a). About 30% of the global electricity generation comes from renewables, 10% from nuclear energy, 23% gas, 36% coal, and 2% oil (IEA 2023a).

The transition to renewable energy and away from fossil fuels for energy generation is necessary to mitigate greenhouse gas emissions and reduce the scale and severity of climate change risks (Aghahosseini et al. 2023; Bogdanov et al. 2021; Jayachandran et al. 2022). The Fourth National Climate Assessment reported that reductions in current emissions can have "modest temperature effects in the near term," but reductions in emissions are "necessary to achieve any long-term objective of preventing warming of any desired magnitude" (Martinich et al. 2018:1351). In the short term, mitigation can help to reduce losses associated with climate change, like perennial sea ice; however, in the longer term, the reduction of greenhouse gas emissions is necessary to avoid crossing critical climate thresholds like marine ice sheet instability and the extremes of global sea-level rise. Taken together, the scientific evidence suggests that decisions in the short term about our energy mix will have serious implications for the effects of climate change over the next century.

Domestic and international actions to support renewable energy

An increasing number of international commitments have been aimed at addressing greenhouse gas emissions and climate change. Under the 2015 Paris Accord, countries agreed to limit warming below 2°C and to stabilize warming to 1.5°C. The temperature thresholds established in the Paris Agreement were set to avoid the most severe impacts of climate change (Martinich et al. 2018). Many parties to the Paris Agreement are working to reduce greenhouse gas emissions by expanding the deployment of renewable energy technology and increasing the energy efficiency of buildings and infrastructure (International Renewable Energy Agency [IRENA] 2022).

Reaching the goals of the Paris Agreement requires that countries transition from fossil fuels to renewable sources of electricity. So far, only a small number of countries have committed to a percentage of renewables in their energy mix, and even fewer have explicit targets for electricity (IRENA 2022). However, China and the European Union (EU) are emerging as two of the largest markets for renewable energy capacity and installation. China has established itself as a leader in renewable energy deployment bolstered by its goal to achieve net-zero by 2060. By 2024,

China is projected to expand its renewable power to represent 55% of the global renewable energy capacity (IEA 2023b). Many factors have supported the growth of European renewable energy technologies – the desire to reduce dependence on Russian natural gas imports, high electricity prices, and national policy support in European markets, especially from Germany, Italy, and the Netherlands (IEA 2023b). In its "Fit for 55" package, the EU has revised its goals to make renewable energy 40% of the energy mix by 2030. The package also outlines goals to reduce greenhouse gas emissions by at least 55% by 2030.

In the United States, the Biden administration has also set ambitious targets for reducing greenhouse gas emissions and increasing the capacity of renewable energy. The administration reentered the Paris Agreement in 2021 with a national commitment to reduce net emissions by 50% to 52% by 2030. The administration announced targets to achieve a carbon-pollution-free power sector by 2035 and a net-zero emissions economy by 2050. Other goals outlined by the administration specific to renewable energy include deploying 30 gigawatts (GW) of offshore wind technology by 2030 and 15 GW of floating offshore wind technology by 2035. The Biden administration's goals are ambitious, given the history of Cape Wind and very limited offshore wind development to date. As of May 2023, there were just 42 megawatts (MW) of installed offshore wind power (Block Island Wind Farm and Coastal Virginia Offshore Wind projects) and 932 MW of offshore wind energy under construction (Vineyard Wind I and South Fork Wind Farm projects).

As the capacity for renewable power is increasing across the United States, the uptake and availability vary across regions (EIA 2023b). There remains significant untapped potential for renewable energy; in 2020, installed renewable energy constituted only 0.2% of the potential energy resources available (Brooks 2022). Even with the increasing capacity of renewable energy and recent political support, there is room for appreciable growth, and the growth in renewables will be contingent on addressing barriers and challenges associated with implementation. Further uptake will also be critical if the United States wishes to achieve its goal of a net-zero emissions economy.

Social acceptance

Although renewable energy now comprises a substantial share of energy generation in some parts of the world, the road to deployment has not been smooth. Social opposition to renewable technologies is widely viewed as an obstacle to this transition. Acknowledging this challenge, scholars have developed *social acceptance* as a subfield of energy social science. An early framework for social acceptance was offered by Wüstenhagen, Wolsink, and Bürer (2007), which identifies three interrelated concepts:

Sociopolitical acceptance is general support for renewable energy and encompasses public opinions regarding energy sources and how policy facilitates or obstructs the adoption of renewables or specific technologies.

Market acceptance reflects financial viability, including whether the costs of energy produced by a technology compare favorably to others or whether investors are willing to back projects.

Community acceptance addresses how people respond to specific projects in the places they live, work, or otherwise value.

This area of research has grown to encompass wide-ranging theories and methods to explain diverse responses to renewable energy policy and infrastructure (Batel 2020). A common theme in the literature has been what is viewed as a conflict between sociopolitical and community acceptance. Most readers will be familiar with the term *NIMBY* as an explanation for community opposition. NIMBY implies that a person or a community supports renewable energy in principle but "not in my backyard." However, social scientists argue that the NIMBY framework is too simplistic, is often used as a pejorative (indicating that people oppose public goods for selfish reasons), and is dismissive of genuine public concerns. In their revisiting of the social gap (a term they coined to describe the discrepancy between high levels of sociopolitical acceptance and the slow deployment of renewable energy projects), Bell and colleagues (2013) acknowledge that NIMBYism exists; however, research has shown that NIMBYism does not explain a significant amount of the public's motivations for opposing projects (Wolsink 2006). Social science has instead offered a wide range of explanations for how the public contributes to the social gap.

Exploring public acceptance of renewable energy

While some members of the public are fundamentally opposed to renewable energy, and others are staunch advocates of any renewable technology, most members of the public hold more nuanced and flexible views. Wolsink (2007) found that public relationships with renewable energy projects follow a U-shaped curve. Generally supportive of renewable energy development in principle, that support tends to drop when a specific project is proposed. This difference may be explained by the different associations people have with the perceived benefits of renewable energy (improved air quality, jobs and economic development, reductions of greenhouse gas emissions) and the potential drawbacks of specific projects (visual impacts, nuisances of construction, threats to local wildlife) (Van der Horst 2007). For example, a local resident may support replacing coal-fired plants with wind energy, but when a project is proposed near a local forest, they worry that the songbirds they enjoy in the spring will be killed by collisions with the turbine blades. Once construction is completed and the project is operational, however, public attitudes tend to rebound to some degree.

A great deal of attention has been given to the visual impacts of renewable energy development. This is, in part, because renewable energy projects – particularly the two most common, wind and solar – tend to have a larger footprint than traditional power plants. Moreover, these projects are often sited in rural or semirural areas, which historically have not hosted electricity generation. In these places, renewable energy development is a significant alteration of the landscape, and residents may view wind turbines and solar panels as industrial invaders in the countryside.

Focusing more on the symbolism than on aesthetics, scholars have sought to understand how public relationships with landscapes drive opposition to renewable energy developments. Two concepts stand out as particularly important: place attachment and fit. Place attachment describes the psychological bond that people have to the places they live, work, and otherwise value (Stedman 2002). The more attached an individual is to the location of a proposed energy project, the more likely they are to oppose it. This is because the project is viewed as a "place disruption"; that is, it threatens the integrity of a person's identity or their ability to participate in the activities they value (Devine-Wright and Howes 2010). Alternatively, the concept of "fit" is related to place meaning (Salak et al. 2021). For example, a natural area can be a place where we spend time with family, a home for wildlife, or a location of natural beauty. When confronted with a new proposed use in a location, people evaluate whether that use is consistent or inconsistent with the meanings they associate with that place. If the new use is consistent with those meanings, we say it "fits."

More broadly, the symbolic nature of renewable energy technologies acts as a filter through which beliefs and attitudes are formed (McLachlan 2009). Research shows that if a person perceives a renewable project as symbolic of progress toward clean energy, they are more supportive of the project. Perceptions of renewable energy systems have also been linked to broader ideologies and guiding psychological principles. Using a common social-psychological framework called values-beliefs-norms (or VBN) for short, research has revealed that altruism (i.e., concern for the welfare of others) is correlated with higher levels of support for renewable energy projects, while greater adherence to traditional values (i.e., attachment to the status quo) predicts less supportive attitudes (Bidwell 2013; 2023).

Beliefs about the fairness of a project are also critical predictors of public attitudes (Firestone et al. 2012; Wolsink 2007). Most of the literature focuses on two broad categories of fairness: procedural and distributive. Distributive fairness (also known as distributive justice) concerns how the benefits and burdens of renewable energy development are allocated. Members of the public may feel that they unfairly bear the negative impacts of a project without commensurate benefits. They may believe, for example, that the natural character of their viewshed is marred by wind turbines while the power generated mostly serves a far-off city. In a broader sense of justice, they may also believe that private companies and their investors are profiting off a public resource.

Procedural fairness (justice) describes beliefs about how decisions are made. One procedural aspect of decision-making for renewable energy development that has received a great deal of attention is the level of community or public engagement in decisions. Generally, the public is interested in knowing that their values and priorities have been considered. Other aspects that contribute to the fairness of decision making include whether the public believes it has been treated with respect and the perceived quality of decisions (i.e., were decisions made based on clear criteria and supported by sound evidence). In our studies of offshore wind energy, for example, we've heard a lot of cynicism from coastal residents that these developments are "a done deal" and that their input isn't taken seriously.

Closing the gap

Research findings about why people support or oppose projects can inform strategies to close the social gap between general support for renewable energy systems and the slow pace of their siting and deployment. The first strategy is to develop technologies and projects that respond to public concerns. This means incorporating public values into technology design. This could include aesthetic considerations but also incorporating features that mitigate harm to wildlife (for example, wind turbines that are less harmful to bats and birds). It could also mean siting projects in landscapes where renewable energy systems have higher levels of perceived "fit" (industrial areas or brownfields, perhaps). Yet, concerns over efficiency and costs, as well as physics and transmission requirements, will always present some limitations to design and siting. Moreover, because beliefs are heavily influenced by underlying values and ideologies, public perceptions regarding potential impacts of renewable energy development may be difficult to change. Most of the literature has focused instead on addressing procedural and distributional aspects of development.

Scholars and practitioners frequently suggest that the engagement of the public and other stakeholders in decisions regarding renewable energy systems can facilitate project success. There are several factors that can vary in how the public participates in decision making, including the scale of the decision (e.g., setting broad energy policy for a state versus permitting of individual projects), the purpose of participation, the openness of the decision process, and the allocation of decision-making authority (Bidwell 2016).

The timing of this engagement also matters. Early engagement improves project support as it reduces suspicions about decision makers and fears that the projects are deliberately secretive (Gross 2007; Yenneti and Day 2015). Engaging the public early also guarantees that community members are included before decisions are made, helping to ensure that the public has a legitimate influence on the project's process and outcome and avoiding the perception that a project is "a done deal." An individual's perception that their voice is being heard in the decision-making processes affects attitudes toward renewable energy projects (Firestone et al. 2012). Ultimately, both the timing and inclusion of the community's voice are important elements of public engagement to improve the acceptability of project outcomes and allow decision makers to consider local knowledge and experiences in their project designs.

One method developers can use to redress imbalances between project benefits and burdens is to provide community benefits. Community benefits are broadly defined as an additional, positive provision for the area and people affected by development projects (Cowell, Bristow, and Munday 2011). There is not one uniform approach taken by governments or developers in providing community benefits. The benefits themselves can include many, varied provisions with both indirect and direct benefits, including monetary streams, electricity discounts, investment in infrastructure (e.g., parks, roads), educational opportunities, social programs, workforce training and development, and joint ownership. The benefits are also distributed through various compensation schemes and can be administered by different actors – local governments, community organizations, or energy developers.

An example of a formal and legally binding benefit scheme is a community benefits agreement. Community benefits agreements are defined as "legally enforceable contracts, signed by community groups and a developer, outlining the community benefits that the developer agrees to provide as a component of the development project" (Gross 2005:9). Community benefits agreements have the potential to identify and address any unequal distribution of project benefits and burdens (Cowell, Bristow, and Munday 2011; Klain et al. 2017). The negotiation of the community benefits agreement can encourage developers to consider the projects' impacts early in the project (Rudolph et al. 2018).

Researchers have found that the provision of community benefits can have mixed effects on social acceptance, depending on how it is implemented (Klain et al. 2017). Community benefits may be perceived as a government or developer bribe, which can backfire by reducing support for the project (Walker, Wiersma, and Bailey 2014). However, the perception of community benefits as bribes can be mediated if conditions of both distributive and procedural fairness are met (van Wijk et al. 2021). To improve procedural justice, negotiations over community benefits provisions can prioritize earlier discussions and involvement of the local community, which may be able to reduce concerns about the developers and their project plans (Aitken 2010). To improve distributive fairness, community needs should be prioritized, and compensation should be beneficial on a collective, community-wide level. For example, a study in the Outer Banks of North Carolina found that residents want community benefits paid by an offshore wind developer to be managed locally, "Because we know what the needs are" (Tyler et al. 2021). Community benefits and compensation are both important elements of renewable energy projects, but careful attention should be paid to the design and method of implementation of these benefit schemes to improve the social acceptance of justice of renewable energy projects.

The case of the Block Island Wind Farm

The United States has been slow to adopt offshore wind energy, compared to many countries in Western Europe and China. Considered a "pilot project," the Block

Island Wind Farm was the first offshore wind energy development in the United States. It is located within the waters of Rhode Island, just 3 miles off the coast of Block Island, a popular, regional summer tourism destination. The project has an electricity production capacity of 30 MW (powering about 17,000 homes). Undersea cables bring the electricity to an island substation; electricity not used on the island is integrated into the mainland grid via a second undersea cable.

As the first offshore wind project to be constructed, the Block Island Wind Farm has attracted a great deal of attention from social scientists. Through a variety of methods, we and our colleagues have investigated public attitudes toward the project, from permitting through operations, seeking to understand its successful development. Overall, inhabitants of the island and mainland, as well as seasonal residents and visitors to the island, exhibited significant support for the project. This support increased over time (Bingaman, Firestone, and Bidwell 2022). Several factors predicted higher support. These included higher levels of altruistic values (Bidwell 2017; 2023) and the symbolic value of the project in representing progress toward "green" energy (Firestone et al. 2018). The attitudes of recreational anglers are even more influenced by this symbolism than by perceived effects on the fish, the fishing experience, or the viewshed (Bidwell, Smythe, and Tyler 2023). One angler explained, "Being someone who is very environmentally conscious, I think that it makes me feel good and proud . . . to see them there generating green energy" (Smythe et al. 2021). A minority of local survey respondents believed that the project does not "fit" the daytime seascape (Russell et al. 2020). Lower support was connected to traditional values and a belief that the project represents "a loss of something intangible, where all you see is the ocean."

Due to the proximity of the turbines to the shore, Block Island residents bear a larger distribution of visual burdens than others. However, the project also included benefits for the residents and community (Klain et al. 2017). Prior to the construction of the wind farm, the island had no physical connection to the mainland, and electricity was produced by burning diesel oil to run generators. The wind farm brought more consistent electricity and eliminated the need for the dirty, noisy generators. Moreover, the community negotiated the inclusion of high-speed fiber-optic Internet access within the cables.

Procedural aspects of the project, however, were a key driver of the development's success. When an offshore energy development was first proposed in 2008, the state of Rhode Island sponsored a 2-year, participatory ocean planning process to explore areas where it could be sited and its potential impacts. This "Ocean Special Area Management Plan" engaged diverse interests in a data-driven deliberation, resulting in the definition of a Renewable Energy Zone where the Block Island Wind Farm was subsequently proposed (Smythe and McCann 2018). One representative of the developer told us,

Just because someone's opposed to the project, doesn't mean their opinion isn't valuable to the project. We've had some folks who have informed the process for us, even though they are in absolute opposition to the process, so every-body's opinion is of value.

(Dwyer and Bidwell 2019)

People like to be heard and empowered. To that end, two procedural aspects out of the permitting process stand out as key to the success of the Block Island Wind Farm: informal interactions and the hiring of a community liaison. Although community members did not want to eliminate formal regulatory requirements for engagement (e.g., comment periods and hearings), they were more appreciative of informal opportunities where they could interact with developers, regulators, and their peers. For the Block Island Wind Farm project, these opportunities were provided through "science fair"–style open houses and meetings between the developers and interest groups.

To further include local voices, the developer also hired a community liaison, a trusted community member who could answer questions and gather feedback, to facilitate these types of informal interactions. In an interview, one Block Island resident said, "I know that there are people who that made all the difference to them, having that community liaison" (Firestone et al. 2020). Involving local, trusted community members made the local population feel respected and included in the development of the Block Island Wind Farm. This inclusion proved essential to the project's success.

Conclusion

Using fossil fuels to generate electricity is a significant contributor to climate change. Yet, our appetite for electricity remains high. Transitioning to renewable sources of electricity could slow our impacts on the climate, but widespread development of wind and solar infrastructure has been delayed by public opposition. Social science research has provided important insights into the causes of a "social gap" in the adoption of renewable sources of electricity. People care deeply about the effects of energy developments on the landscapes they value. Perceptions of the fairness of projects (how burdens and benefits are distributed, as well as how decisions are made) also affect public attitudes. The success of the Block Island Wind Farm offers some insights into how to engage local residents to build renewable energy systems. Specifically, residents need to be involved from the beginning – listened to and have their concerns addressed. Instrumental to Block Island's success was the use of community liaisons to facilitate informal communication about the project. These efforts increased residents' trust in the process and support for the wind farm. This case study points to the need for more meaningful engagement of the public in planning and decision making for renewable energy policies, specific projects, and how benefits are provided to communities.

References

Aghahosseini, Arman, Solomon, Christian Breyer, Thomas Pregger, Sonja Simon, Peter Strachan and Arnulf Jäger-Waldau. 2023. "Energy System Transition Pathways to Meet the Global Electricity Demand for Ambitious Climate Targets and Cost Competitiveness." *Applied Energy* 331: 120401. https://doi.org/10.1016/j.apenergy.2022.120401

Aitken, Mhairi. 2010. "Wind Power and Community Benefits: Challenges and Opportunities." *Energy Policy* 38(10): 6066–6075. https://doi.org/10.1016/j.enpol.2010.05.062

Batel, Susana. 2020. "Research on the Social Acceptance of Renewable Energy Technologies: Past, Present and Future." *Energy Research & Social Science* 68: 101544.

Bell, Derek, Tim Gray, Claire Haggett and Joanne Swaffield. 2013. "Re-Visiting the 'Social Gap': Public Opinion and Relations of Power in the Local Politics of Wind Energy." *Environmental Politics* 22(1): 115–135.

Bidwell, David. 2013. "The Role of Values in Public Beliefs and Attitudes Towards Commercial Wind Energy." *Energy Policy* 58: 189–199.

Bidwell, David. 2016. "Thinking Through Participation in Renewable Energy Decisions." *Nature Energy* 1: 16051.

Bidwell, David. 2017. "Ocean Beliefs and Support for an Offshore Wind Energy Project." *Ocean & Coastal Management* 146: 99–108.

Bidwell, David. 2023. "Tourists Are People Too: Nonresidents' Values, Beliefs, and Acceptance of a Nearshore Wind Farm." *Energy Policy* 173: 113365.

Bidwell, David, Tiffany Smythe and Grant Tyler. 2023. "Anglers' Support for an Offshore Wind Farm: Fishing Effects or Clean Energy Symbolism." *Marine Policy* 151: 105568. https://doi.org/10.1016/j.marpol.2023.105568

Bingaman, Samantha, Jeremy Firestone and David Bidwell. 2022. "Winds of Change: Examining Attitude Shifts Regarding an Offshore Wind Project." *Journal of Environmental Policy & Planning* 1–19. https://doi.org/10.1080/1523908X.2022.2078290

Bogdanov, Dmitrii, Manish Ram, Arman Aghahosseini, Ashish Gulagi, Ayobami Solomon Oyewo, Michael Child, Upeksha Caldera, Kristina Sadovskaia, Javier Farfan, Larissa De Souza Noel Simas Barbosa, Mahdi Fasihi, Siavash Khalili, Thure Traber and Christian Breyer. 2021. "Low-Cost Renewable Electricity as the Key Driver of the Global Energy Transition Towards Sustainability." *Energy* 227: 120467. https://doi.org/10.1016/j.energy.2021.120467

Brooks, Adria. 2022. "Renewable Energy Resource Assessment Information for the United States." US Department of Energy. Accessed December 12, 2023. www.energy.gov/sites/default/files/2022-03/Renewable%20Energy%20Resource%20Assessment%20Information%20for%20the%20United%20States.pdf

Cowell, Richard, Gill Bristow and Max Munday. 2011. "Acceptance, Acceptability and Environmental Justice: The Role of Community Benefits in Wind Energy Development." *Journal of Environmental Planning and Management* 54(4): 539–557.

Devine-Wright, Patrick and Yuko Howes. 2010. "Disruption to Place Attachment and the Protection of Restorative Environments: A Wind Energy Case Study." *Journal of Environmental Psychology* 30(3): 271–280.

Dwyer, Joseph and David Bidwell. 2019. "Chains of Trust: Energy Justice, Public Engagement, and the First Offshore Wind Farm in the United States." *Energy Research & Social Science* 47: 166–176.

Energy Information Administration. 2023a. "U.S. Energy Facts Explained." Accessed December 12, 2023. www.eia.gov/energyexplained/us-energy-facts/data-and-statistics.php

Energy Information Administration. 2023b. "Annual Energy Outlook 2023." Accessed December 12, 2023. www.eia.gov/outlooks/aeo/narrative/

Environmental Protection Agency. 2023a. "Sources of Greenhouse Gas Emissions." Accessed December 12, 2023. www.epa.gov/ghgemissions/sources-greenhouse-gas-emissions#t1fn1

Environmental Protection Agency. 2023b. "Overview of Greenhouse Gases." Accessed December 12, 2023. www.epa.gov/ghgemissions/overview-greenhouse-gases

Firestone, Jeremy, David Bidwell, Meryl Gardner and Lauren Knapp. 2018. "Wind in the Sails or Choppy Seas? People-Place Relations, Aesthetics and Public Support for the United States' First Offshore Wind Project." *Energy Research & Social Science* 40: 232–243.

Firestone, Jeremy, Christine Hirt, David Bidwell, Meryl Gardner and Joseph Dwyer. 2020. "Faring Well in Offshore Wind Power Siting? Trust, Engagement and Process Fairness in the United States." *Energy Research & Social Science* 62: 101393.

Firestone, Jeremy, Willett Kempton, Meredith Blaydes Lilley and Kateryna Samoteskul. 2012. "Public Acceptance of Offshore Wind Power: Does Perceived Fairness of Process Matter?" *Journal of Environmental Planning and Management* 55(10): 1387–1402.

Gross, Catherine. 2007. "Community Perspectives of Wind Energy in Australia: The Application of a Justice and Community Fairness Framework to Increase Social Acceptance." *Energy Policy* 35(5): 2727–2736.

Gross, Julian. 2005. "Community Benefits Agreements: Making Development Agreements Accountable." *Good Jobs First.* Accessed December 12, 2023. www.goodjobsfirst.org/wp-content/uploads/docs/pdf/cba2005final.pdf

International Energy Agency. 2023a. "World Energy Outlook 2022". Accessed December 12, 2023. www.iea.org/reports/world-energy-outlook-2022

International Energy Agency. 2023b. "Renewable Energy Market Update–June 2023." Accessed December 12, 2023. www.iea.org/reports/renewable-energy-market-update-june-2023

International Renewable Energy Agency. 2022. "NDCs and Renewable Energy Targets in 2021: Are We on the Right Path to a Climate-Safe Future?" Accessed December 12, 2023. www.irena.org//media/Files/IRENA/Agency/Publication/2022/Jan/IRENA_NDCs_RE_Targets_2022.pdf?rev=621e4518d48f4d58baed92e8eda3f556

Jayachandran, M., Ranjith Kumar Gatla, K. Prasada Rao, Gundala Srinivasa Rao, Salisu Mohammed, Ahmad H. Milyani, Abdullah Ahmed Azhari, C. Kalaiarasy and S. Geetha. 2022. "Challenges in Achieving Sustainable Development Goal 7: Affordable and Clean Energy in Light of Nascent Technologies." *Sustainable Energy Technologies and Assessments* 53: 102692. https://doi.org/10.1016/j.seta.2022.102692

Klain, Sarah, Terre Satterfield, Suzanne MacDonald, Nicholas Battista and Kai M A. Chan. 2017. "Will Communities 'Open-up' to Offshore Wind? Lessons Learned from New England Islands in the United States." *Energy Research & Social Science* 34: 13–26.

Martinich, Jeremy., Benjamin DeAngelo, Delvane Diaz, Brenda Ekwurzel, Guido Franco, Carla Frisch, James McFarland and Brian O'Neill. 2018. "Reducing Risks Through Emissions Mitigation." In *Fourth National Climate Assessment*, 1346–1386. Washington, DC: US Global Change Research Program. https//doi.org/10.7930/NCA4.2018.CH29

McLachlan, Carly. 2009. "'You Don't Do a Chemistry Experiment in Your Best China': Symbolic Interpretations of Place and Technology in a Wave Energy Case." *Energy Policy* 37(12): 5342–5350.

Rudolph, David, Claire Haggett and Mhairi Aitken. 2018. "Community Benefits from Offshore Renewables: The Relationship Between Different Understandings of Impact, Community, and Benefit." *Environment and Planning C: Politics and Space* 36(1): 92–117.

Russell, Aaron, Jeremy Firestone, David Bidwell and Meryl Gardner. 2020. "Place Meaning and Consistency with Offshore Wind: An Island and Coastal Tale." *Renewable and Sustainable Energy Reviews* 132: 110044.

Salak, Boris, Kreg Lindberg, Felix Kienast and Marcel Hunziker. 2021. "How Landscape-Technology Fit Affects Public Evaluations of Renewable Energy Infrastructure Scenarios. A Hybrid Choice Model." *Renewable and Sustainable Energy Reviews* 143: 110896. https://doi.org/10.1016/j.rser.2021.110896

Smythe, Tiffany, David Bidwell and Grant Tyler. 2021. "Optimistic with Reservations: The Impacts of the United States' First Offshore Wind Farm on the Recreational Fishing Experience." *Marine Policy* 127: 104440.

Smythe, Tiffany C. and Jennifer McCann. 2018. "Lessons Learned in Marine Governance: Case Studies of Marine Spatial Planning Practice in the US." *Marine Policy* 94: 227–237.

Stedman, Richard C. 2002. "Toward a Social Psychology of Place: Predicting Behavior from Place-Based Cognitions, Attitude, and Identity." *Environment and Behavior* 34(5): 561–81.

Tyler, Grant, David Bidwell, Tiffany Smythe and Simona Trandafir. 2021. "Preferences for Community Benefits for Offshore Wind Development Projects: A Case Study of the Outer Banks of North Carolina, U.S." *Journal of Environmental Policy & Planning* 24(1): 39–55. https://doi.org/10.1080/1523908X.2021.1940896

Van der Horst, D. 2007. "NIMBY or Not? Exploring the Relevance of Location and the Politics of Voiced Opinions in Renewable Energy Siting Controversies." *Energy Policy* 35(5): 2705–2714.

van Wijk, Josef, Itay Fischhendler, Gillad Rosen and Lior Herman. 2021. "Penny Wise or Pound Foolish? Compensation Schemes and the Attainment of Community Acceptance in Renewable Energy." *Energy Research & Social Science* 81: 102260. https://doi.org/10.1016/j.erss.2021.102260

Walker, Benjamin, Bouke Wiersma and Etienne Bailey. 2014. "Community Benefits, Framing and the Social Acceptance of Offshore Wind Farms: An Experimental Study in England." *Energy Research & Social Science* 3: 46–54.

Wolsink, Maarten. 2006. "Invalid Theory Impedes Our Understanding: A Critique on the Persistence of the Language of NIMBY." *Transactions of the Institute of British Geographers* 31(1): 85–91.

Wolsink, Maarten. 2007. "Wind Power Implementation: The Nature of Public Attitudes: Equity and Fairness Instead of 'Backyard Motives.'" *Renewable and Sustainable Energy Reviews* 11(6): 1188–1207.

Wüstenhagen, Rolf, Maarten Wolsink and Mary Jean Bürer. 2007. "Social Acceptance of Renewable Energy Innovation: An Introduction to the Concept." *Energy Policy* 35(5): 2683–2691.

Yenneti, Komali and Rosie Day. 2015. "Procedural (in)Justice in the Implementation of Solar Energy: The Case of Charanaka Solar Park, Gujarat, India." *Energy Policy* 86: 664–673.

16

FUTURES BORN OF THE PAST AND PRESENT

Building transitions as collaborative projects of justice

Tristan Partridge and Javiera Barandiarán

The problem: Energy transitions risk replicating the exploitative relations and growth imperatives of the fossil fuel economy, minimizing any positive climate impacts.

The solution: More expansive approaches: reframing transitions as sociopolitical processes where collective action, centered on consent and reciprocity, is allowed to thrive; and supporting diverse initiatives that are already underway, particularly those that reframe climate-focused transitions as opportunities to radically redefine, redistribute, and relocate power.

Where this has worked: Radical rearrangements of social, economic, and environmental relations toward justice objectives already working include the Transition Towns movement in the United Kingdom; farmer-run solar cooperatives in India; land reform–led energy initiatives in Scotland; localized solar generation and "energy reparations" initiatives in New Orleans; "energy democracy" activists in New York City and elsewhere engaged with the Solarize model of community energy.

This chapter draws on our work on the social and environmental effects of changes to energy systems at local, regional, and global levels. We examine how different groups think about and build multiple forms of energy transition. We highlight how there is no single "right" way to organize energy nor to change it from its current reliance on fossil fuels to new energy sources.

DOI: 10.4324/9781003437345-22

Some of the broader goals of energy transitions are clearly established, such as aiming to build energy systems that

- minimize environmental harms,
- address rather than exacerbate social inequalities, and
- reduce use of fossil fuels.

However, the specific dynamics of transitions remain up for debate. Specifically, *who designs, constructs, and controls processes of transition are open questions*. Social movements are often forced to compete against companies and governments involved in energy transitions who see themselves as the parties in charge of organizing these goals as a single transition focused on profits (Partridge 2022).

In our work, we focus on transitions as opportunities to think differently about or dismantle dominant political, economic, and social institutions (Howe and Boyer 2016). This chapter therefore highlights diverse forms of resistance to the ways those in power are co-opting energy transitions. As social scientist Larry Lohman has noted, those advocating for reorganizing energy systems need to be critical about the fossil fuel industry's goals:

> As their reports and advertisements frankly reveal, most oil companies, banks and industrial corporations see energy transition not as a process that will replace fossil fuels, but as a process that will supplement them . . . as a way of delivering better returns on investments that, at bottom, will go on being organized around oil, coal, and gas. For them, an energy transition is a way of diversifying and intensifying the same type of labour exploitation that fossil capitalism made universal. Climate activists need to be careful lest their own advocacy of "transition" merely plays into this dynamic.
>
> *(Lohmann 2015:6)*

Countless cases globally show how much industrial-scale "renewable" energy development exacerbates exploitative relationships with the earth and with workers (Dunlap 2019).[1] Market-led and consumption-based transition initiatives prioritize the bottom line, which threatens to deepen rather than alleviate environmental injustices experienced by the world's most marginalized populations. In China, for example, farmland has been cleared for new industrial parks to manufacture solar panels, leaving many farmers with no choice but to accept exhausting, poorly paid manufacturing jobs (Chen 2013). Consumption-led transitions also lead to an ongoing expansion of electricity production and resource use (Kallis et al. 2020; York and Bell 2019).

Yet, at the same time, a rising global movement is mobilizing against neoliberal (see Chapter 8 for an explanation of neoliberalism) and extractive energy transitions – highlighting the analytical goal of re-politicizing energy debates that are increasingly dominated by apolitical, technocratic, and growth-oriented economic frames and objectives (Avila-Calero 2017). By paying attention to how histories of both resource extraction and political mobilization shape the immediate potential for building less destructive energy systems, we examine what actions generate *energy transitions as collaborative projects of justice*, that is, as "socio-political processes where collective action, centered on consent and reciprocity, is allowed to thrive" (Partridge 2022:47). In this examination, we emphasize the importance and influence of place-based histories. We consider transitions to be always in relation to theories of change and visions for the future that help build social movements – particularly when articulating and acting toward shared visions that refuse the imitation and replication of elite values (Fanon 1963; Maeckelbergh 2016).

Our title here echoes a widely read text from 1913 that emphasizes how any contemporary action or innovation builds upon societal precedents and is impossible without the work of global majorities. The very conditions of possibility for enacting an intentional energy transition (or transition of any kind) are shaped by prior histories of many different kinds, including distinct sociopolitical and geographical histories of collective organization, industrial production, social mobilization, and resource extraction, among others. The text in question by Kropotkin reads:

> Millions of human beings have labored to create this civilization . . . Other millions, scattered through the globe, labor to maintain it. Without them nothing would be left in fifty years but ruins. There is not even a thought, or an invention, which is not common property, born of the past and the present . . . Thousands of writers, of poets, of scholars, have labored to increase knowledge, to dissipate error, and to create that atmosphere of scientific thought . . . They have been upheld and nourished through life, both physically and mentally, by legions of workers and craftsmen of all sorts. They have drawn their motive force from the environment.
>
> *(Kropotkin 1995:15)*

Our goal here is to apply these insights in reconsidering how energy transitions are framed and discussed in the face of unfolding climate crises – underlining how visions for the future are 1) shaped by the past and 2) emergent within ongoing, place-specific struggles for social and environmental justice.

The past is still with us

There are many global practices and concepts that explicitly orientate visions for the future around a deepened understanding of the past. One example with roots in West Africa is embodiment of *sankofa*, a word used in Akan wisdom and philosophical life that is conventionally translated into English as "go back and fetch it," "return to your past," or "it is not taboo to go back and retrieve what you have forgotten or lost" (Temple 2010:127). The idea inspires multiple forms of resistance. Recovery, repair, and reclaiming control over what is yet to happen begins with reclaiming what has been lost, or stolen, before now. Anishinaabe writer Patty Krawec (2022) specifically calls for collective action today to be rooted in "unforgetting" histories of resistance and struggle – particularly when renewed relations among and between groups, including human and nonhumans, serve to help disrupt histories of violence and denial of rights. That said, reading such practices as a form of traditionalism, or as a *return* to an imagined past, is a mistake. Rather, this orientation is closer to what Ariella Azoulay explores as unlearning imperialism – that is, making explicit the ways in which our thoughts, actions, and modes of existence are shaped by processes of imperialism, then refusing those forms of violence, and committing instead to processes of repair (Azoulay 2019). In other words, collective futures are rooted in the present: they emerge in the organizing, remembering, resisting, and mobilizing taking place right now.

This re-presenting of the future is an orientation found also in other political movements, albeit in somewhat different forms. Stuart Hall maintained that committing to the construction of a new political will must be grounded in analyses of the present that are neither "rote" nor "celebratory" and which attend to "things as they are, without illusions or false hopes" as the basis for actions that help us "transcend the present" (Hall 1988:13–14).

The grassroots collective of co-researchers, Colectivo Situaciones, incorporates a similar ethos into their processes of militant/activist research. In their practices, they enact the possibility of liberation in the present by working from

> the power (*potencia*) of what is and not of what "ought to be" . . . Research militancy does not extract its commitment from a model of the future, but from a search for power (*potencia*) in the present. That is why the most serious fight is against the a priori, against predefined schemes.
>
> *(Colectivo Situaciones 2007:84)*

Elsewhere, this approach further underlines the power of potential, lying in uncertainty, possibility, and openness to alternatives: "it's not reasons that make revolutions, it's bodies" (The Invisible Committee 2017:7). One way to engage with such approaches is to identify and challenge how histories of suffering and exploitation curtail contemporary action (Barandiarán 2022; Frazier 2007). Meaningful transitions need to work to dismantle and overcome those curtailments.

Specific cases directly address the role of energy in such transitions. Andrew Curley opens his book, *Carbon Sovereignty*, with the observation that

> energy transition is not simply an ideal for the future; it is also empirically a series of past events. Included in energy transition are the marginalization of tribal places, the expansion of unsustainable cities, and the slow violence of toxic spaces. Transition is not just a political rhetoric or rallying point; it is made violent by its implementation on already existing colonial landscapes.
>
> *(Curley 2023:5)*

Curley examines the clean energy campaign of Diné environmental activists that centers *transitions* around Diné lifeways, land, and alternative visions of survivance (Curley 2023:28). The latter is a term that Gerald Vizenor describes as "an active sense of presence over historical absence, deracination, and oblivion" (Vizenor 2008:1). The complexities of Diné histories of coal extraction and emergent energy politics challenge any singular understanding of "transition." Amid debates about what kinds of politics might better sustain Native economies, ecologies, and communities, significant work has been done to articulate "an environmental politics in which transition *away* from large-scale extraction of natural resources can go hand-in-hand with transition *toward* a more autonomous, self-reliant, sovereignty Native Nation" (Powell 2017:223). In this sense, transitions as collaborative projects of justice require energy projects, energy relations, and infrastructure development to be designed so that they further collective goals, even while those objectives (and the specific means for achieving them) may remain contested.

A further consequence of this orientation toward transitions is that collective work toward building futures involves *both* looking forward and looking back: not only anticipating the diverse desired futures of transitions and climate resiliency but also addressing ongoing historical injustices and scrutinizing their effects. For example, elsewhere we have argued that to achieve the dual goals of minimizing global pollution and meeting diverse demands for environmental justice, energy transitions need to also involve the safe decommissioning of older energy infrastructures and management of their toxic legacies (Partridge et al. 2023). For impacted communities, energy transitions will only be effective when, for example, the air and water contamination caused by nearby leaking oil wells has been addressed. The scale of this problem globally continues to grow – and many communities are put at increased risk as the owners of oil and gas wells go bankrupt or become elusive now that the wells are no longer profitable. Even contained and seemingly well-run decommissioning operations, like that at Rincon Island off the central California coast, leave communities with the conundrum of what comes next (Barandiarán et al. 2022; CREW 2023). Expanding the temporal frame of analysis around transitions emphasizes that time does not necessarily pass in simple, linear ways; responses to climate change must address processes that can be slow, fast, sudden, or all at the same time. The specific needs of societies to safely and equitably decommission multiple forms of old or abandoned fossil fuel infrastructure is a case in point. Both locally and globally, we will have to live with the polluting legacies of "legacy" fuels for countless years to come.

A word of caution here. The concept of transition implies epochs, or very long periods of time, and different rates of change from one epoch to another (Curley 2023). Particularly in places shaped by histories of resource extraction, demands for an end to ecological disruption may envision transition as a "post-extractivist rupture" – a defiant assertion that fossil fuels are not an inevitable basis for human life – even while the process(es) of enacting such a transition take time (Powell 2017). Climate change is routinely framed in terms of urgency and the immediate need to achieve a rupture with fossil fuels. This urgency, however, is also used to justify energy policies that may help meet short-term targets to reduce emissions but do nothing to address underlying systemic injustices that require long-term social change (Barandiarán 2019; Partridge et al. 2018). Much transitions discourse is similarly troubling, focusing *only* on the present while overlooking histories of struggle and change. This results in proposals that diagnose, at a particular moment in time, social and environmental faults with certain components of energy systems, yet put forward only limited modifications in response – all framed in the language of crises, problems, and solutions (Partridge 2022). Too narrow a focus on the present means a risk of replicating ahistorical analyses and policies.[2] This is analytical "presentism": focusing on the urgency of injustice without also demanding the radical, long-term, and structural transformations that would serve to dismantle colonial power and end violence against marginalized groups (Whyte 2021). As with the language of urgency, the language of crisis risks obscuring from view underlying causes of social and environmental injustice.

When a discourse of crisis is amplified (typically followed by innovation as the best response), analytical presentism fails to recognize how past experiences of violence shape differing degrees of power for enacting more autonomous and just social and energy systems (Whyte 2021). No amount of urgent action, implemented today, can change the fact that Indigenous peoples have inhabited a "postapocalyptic world" since the invasions of colonialism began – an epoch also characterized by resistance, kin-making, and resurgence (Dillon 2012; Gross 2014; Partridge 2022). For many people, it is likely already too late to avoid the dangerous effects of climate change, but the struggles for justice continue (Whyte 2020). Yet much energy research writing on transitions uncritically adopts an institutional perspective on social change. Such authors assume that powerful actors will do the right thing when presented with sufficient information (Healy and Barry 2017). Or they assume, again falsely, that extant political and judicial systems can deliver justice for Indigenous and marginalized groups even when such systems "routinely denigrate and discriminate against those same people and communities" (Partridge 2022:56). By contrast, transitions as collaborative projects of justice build on processes of unforgetting and repair, building presence and survivance, and foregrounding the role of diverse resistance movements who are already engaged in struggles for survival.

Support for the work that is already being done

Along with (re)framing transitions to cover more complicated temporal frames, it is necessary to support diverse place-based struggles toward sustainability – typically mobilized in response to historical dispossession – rather than hoping that new technologies alone will provide solutions or expecting equitable results from reformist rearrangements of resource-intensive systems of production and consumption. Technologies may of course play a role but tend to have more long-lasting positive effects when they are designed and used in response to specific local needs. For example, in our own work in Dhundi village in India, a community-led and nongovernmental organization–financed project established a "solar cooperative that produces Solar Power as a Remunerative Crop" as a holistic energy initiative: combining localized solar energy generation with groundwater management and income support for farmers (by replacing diesel pumps with solar power and making surplus energy available for sale to the state grid, thus incentivizing a reduction in overall water extraction) (Partridge 2022). What began as a pilot project has built on its success and continues to grow its community of solar farmer participants.

Still, there are no silver bullet solutions. Instead, the reframing of transitions we are describing places the focus on how different communities, at different scales, can reassess and redesign relations with the fossil fuel industry by challenging the underlying imperatives of contemporary extractive economies (Partridge 2017b). By redesigning how the products and practices of fossil fuel extraction are avoided or integrated into everyday life, future-oriented action can be primarily rooted in the resources and realities immediately at hand – actions which, in many contexts, begin with communities reclaiming control over land and political processes (Kenrick 2011; Partridge 2017b). In certain parts of Europe, these strategies have been fundamental to building awareness around the transition concept.

For almost 20 years, the Transition Towns movement has encouraged communities to creatively explore strategies for building resilience, tailoring coordinated actions to place-specific social dynamics and ecosystem relations (Hopkins 2008; Partridge 2022). The movement grew from a collective, pedagogical permaculture project that was published in 2005: "Kinsale 2021: an Energy Descent Action Plan" (ibid.). As cited elsewhere (Partridge 2022:52), suggested principles to support "energy descent" include

1) localizing production of food, energy, and building materials wherever possible;
2) designing projects and enterprises that are low carbon with regard to both inputs and outputs;
3) bringing assets (such as land, businesses, energy generation, buildings) into community ownership;
4) creating a vision of abundance for the future while recognizing that our world is one in which credit, resources, and the materials that support energy systems are finite; and
5) endorsing business models that are not purely for personal profit (such as social enterprises and cooperatives) (Hopkins 2008; 2019).

The breadth of focus of these proposals reflects how differently they might be enacted and realized in diverse global contexts. Each approach plays a part in supporting locales and communities in their efforts to endure, respond to, and hopefully survive the effects of global climate change. Addressing and reducing overall energy use remains central to these efforts.

The many diverse goals of transitions as collaborative projects of justice are thus based in radical reimaginings of place. This means reconnecting energy transitions with a politics of conviviality (Illich 1973), with the politics of degrowth and radical global redistribution (Kallis et al. 2020), as well as abolition and gender justice (Heynen and Ybarra 2021). Abolition ecologies draw on Ruth Wilson Gilmore's *abolition geography* and the foundational premise that "freedom is a place" (Gilmore 2017:227). Abolition, understood here as "the destruction of racial regimes and racial capitalism," entails

> not only the end of racial slavery, racial segregation, and racism, but also the abolition of a capitalist order that has always been racial, and that not only extracts life from Black bodies, but dehumanizes all workers while colonizing indigenous lands and incarcerating surplus bodies
>
> *(Johnson and Lubin 2017:12)*

Place-making processes, thus engaged with abolition, are about more than resource management and localization (both of which are practices that can be fundamental to rebuilding collective political power, as is the case with many commons-based initiatives) – they are processes that amass and align collective powers to end oppressions experienced by specific communities in specific locations.

These are powerful actions that urge a deeper reframing of climate-focused transitions as opportunities to radically redefine, redistribute, and relocate power. Drawing on abolition ecologies, Black radical thought, and oral histories in Washington, DC, Ranganathan and Bratman (2021) describe abolitionist climate justice as a radical reimagining and reorientation of resiliency not toward adaptation to "future external threats," but instead as an ethics of care, mutuality, and healing – an approach to climate justice which analytically centers ongoing historical oppressions and the "intersectional drivers of precarity and trauma" experienced by people who are most exposed to climate change impacts (Ranganathan and Bratman 2021).

Affected communities are not defined by the violence they suffer, however: abolition ecologies call for "attention to radical place-making and the land, air and water based environments within which places are made . . . [demanding] attention to the ways that coalitional land-based politics dismantle oppressive institutions" (Heynen and Ybarra 2021:21). Again, this connects to the immediacy of transitions as collaborative projects of justice: often the most pressing need is more substantial support for work that is already being done, more opportunities for acts of resistance to achieve their goals.

Histories of collective organizing

In New Orleans, coalition movements have made the case for localized solar projects as a form of "energy reparations" (Luke and Heynen 2020). This is a process of "relocating control of energy systems, and the multiple forms of power associated with that control, within low-income and Black communities" both as an act of reparation and as "a forward-looking strategy for dismantling the processes of slavery, patriarchy, imperialism, and genocide that fuel a status quo defined by 'petro-racial capitalism' (Luke and Heynen 2020)" (Partridge 2022:32). The potential for redesigned energy systems and energy relations to contribute to movements for social justice is here brought into the center of transitions thinking.

In New York City, Myles Lennon (2021) documents the work of "energy democracy" activists engaged with the Solarize model of expanding residential solar energy generation through local employment and in support of community-based economies, and also as a way of reclaiming power within accountable, collective organizations. Similarly, this work places energy system change firmly within a social context – specifically one that seeks to address ongoing histories of exploitation and exclusion. These activists

> caution against a renewable-powered world where electricity continues to be controlled by unaccountable investor-owned utilities who focus on maximizing their profits on the backs of rate-payers in communities of color; where fossil fuel corporations invest heavily in industrial-scale solar and wind farms to abet – not attenuate – their extractive practices; where renewable energy developers occupy and steal indigenous lands to power the infrastructure of multinational corporations; and where communities of color continue to supply undervalued labor to the energy generation industry and see none of the financial benefits of a broad energy transition.
>
> *(Lennon 2021:4)*

By making these multiple dimensions of energy transitions explicit – and with a focus of collective activism – addressing overlapping inequalities becomes an inseparable goal of energy systems change.

Crucial to the project goals of the Dhundi solar farmers' cooperative, mentioned earlier – "to reconfigure our power economy, our groundwater economy, and our agrarian livelihoods" (Shah et al. 2017:15) – is a local history of success with cooperative enterprises. Dhundi is located in Gujarat state, home to one of the world's largest agricultural cooperatives, with the effect that "cooperative organizing, joint ownership, collective earning, and pooling resources for mutual economic benefit – these are practices that, rather than being dismissed or viewed with suspicion, are widely seen as positive and productive modes of organizing" (Partridge 2022:141). As the number of decentralized solar projects globally continues to grow, the Dhundi solar farmers' cooperative underlines a fundamental consideration for such forms of energy transition: localization projects largely depend on community labor, resources, relationships, and cooperation to succeed. And this success is greatly facilitated in communities that have prior experiences of collaboration, cooperation, and histories of collective organizing.

Today, a community of just over 100 people on the small isle of Eigg, 5 miles off the West Coast of Scotland, enjoy the benefits of a community-owned, community-run, island-specific micro-grid based on a mix of renewable energy (with backup diesel generators). Built in 2008, the micro-grid radically altered residents' relationships with fossil fuels, replacing previous energy provisioning that saw each individual property dependent on noisy, polluting, expensive, and inefficient household generators, dependent on diesel shipped over from the mainland. Crucially, in June 1997, the residents of Eigg had coordinated with other communities and land-reform activists and were successful in launching a community "buy-out" of the island – reclaiming from a single landlord collective control of the land, planning, development, and infrastructure governance while developing accountable decision-making processes. The decarbonization of electricity supplies on Eigg is thus bound up with these shared acts of collective reclamation. As a transition initiative, the micro-grid project was enabled by three mutually reinforcing factors:

1) Directly democratic and accountable decision-making
2) Community control of land and other resources
3) Collaborative organizing developed through engagement with similar communities (Partridge 2022:154)

This is another context where energy transition initiatives are inseparable from collective efforts to address historical inequalities.

On one hand, rooting thinking and action about energy transitions in place-based social projects can be empowering – fostering other related forms of resiliency and supporting work toward other related justice objectives. On the other, this orientation also exposes how much of the actual work and responsibility for creating meaningful change continues to fall to people outside the most recognized positions of power and authority.

The place-specific priorities, needs, and dynamics of any effective "solutions" to the climate crisis will vary across global and local contexts. What all such efforts share in common, as illustrated in the examples presented earlier, is a commitment to reinforcing deeply connected, reciprocal, and consensual forms of interrelation and production while also dismantling systems that depend on exploitation and exclusion (Partridge 2017a). This is the basis of (re)framing transitions as collaborative projects of justice.

Yet, as the cases described here also show, such renewed projects of collectivity – whether a solar farm, a micro-grid, or a relocalization program of the kinds seen in Transition Towns – largely rely on voluntary, precarious, high-risk, or at-risk forms of work. As with other coordinated efforts to meaningfully address social and environmental injustices, the successful implementation of revalorized or reimagined political visions at any scale ultimately depends significantly on the labor, cooperation, and social resources of particular communities – often rural, low-income, Indigenous, communities of color, or other groups most acutely affected by the injustices being addressed (Partridge 2018). Thus, for "transitions" to be meaningful, there has to be explicit, direct, and sustained forms of support provided to groups whose labor and collective efforts are already underpinning and enabling changes to energy systems.

Conclusion

Even while certain populist leaders and other groups funded by the fossil fuel industry may continue to deny or denounce this reality, a transition – of sorts – toward increased renewable energy generation is already underway globally. As we have seen, however, there remain ongoing struggles to ensure that these processes serve the interests and immediate needs of marginalized communities – designing transition processes to not only halt harm but also to reverse historical processes of violence and exploitation. The need to support these struggles is only intensified by global inequalities and the fact that most of those suffering the most acute effects of climate change are those with the fewest resources or opportunities to protect themselves and their communities.

Already, there are branches of the extractive industries that function in a somewhat "zombie" state: effectively dead (lacking solvency or any kind of long-term future) yet outwardly recognizable as an industry and still able to function in ways that resemble their former, more vibrant, condition. As Macey and Salovaara point out (2019:879), as of 2019 almost half of all the coal produced in the United States was being mined by companies that had gone bankrupt. One result of this, seen also across the US oil industry (and reflecting other "zombie" tendencies), is that companies facing their impending demise begin to lash out – through actions that compound environmental damage, disruption, and risk (Partridge et al. 2023). That is, when faced with increasingly uncertain futures, extractive companies act to evade environmental responsibilities through deliberately opaque strategies of corporate restructuring, bankruptcies, shell company registrations, and other methods of "offloading" liabilities (Partridge et al. 2023; Grubert 2020). Again, the effect is that extant social and environmental inequalities are deepened: profits and returns to shareholders are prioritized and the costs of unemployment and industrial cleanup are shifted onto the public sector. All the while, environmental harm and toxification are intensified within already affected communities.

Reframing transitions as collaborative projects of justice creates a broad scope for radically reimagining the potential outcomes of energy transitions. This means always linking transition policies and designs to transformative actions that further support ongoing efforts to dismantle colonial power and end violence against marginalized groups (Whyte 2021). Collaborative projects of justice build on processes of unforgetting and repair, offering support for environmental and climate justice struggles that are already being mobilized – and recognizing how global collective futures are emergent within those mobilizations as well as the acts of remembering and resisting that they incorporate.

Notes

1 Also, appearing with increasing regularity in the media, are reports on how "meaningless" (or misleading, or destructive) climate policies or net-zero pledges by fossil fuel companies are. An example from June 2023 is: www.reuters.com/sustainability/fossil-fuel-company-net-zero-plans-largely-meaningless-report-2023-06-11/
2 For examples of efforts to avoid ahistorical analyses of energy futures, see the special issue of *Media + Environment* on Energy Justice in Global Perspective (Barandiarán et al. 2022).

References

Avila-Calero, Sofía. 2017. "Contesting Energy Transitions: Wind Power and Conflicts in the Isthmus of Tehuantepec." *Journal of Political Ecology* 24: 992–1012.

Azoulay, Ariella. 2019. *Potential History: Unlearning Imperialism*. London: Verso.

Barandiarán, Javiera. 2019. "Lithium and Development Imaginaries in Chile, Argentina and Bolivia." *World Development* 113: 381–391. https://doi.org/10.1016/j.worlddev.2018.09.019

Barandiarán, Javiera. 2022. "Lithium Futures in Chile, Argentina, and the United States." March 10, The American Academy in Berlin.

Barandiarán, Javiera, Mona Damluji, Stephan Miescher, David N. Pellow and Janet Walker. 2022. "Energy Justice in Global Perspective: An Introduction." *Media+Environment* 4(1). https://doi.org/10.1525/001c.37073

Chen, Jia-Ching. 2013. "Sustainable Territories: Rural Dispossession, Land Enclosures and the Construction of Environmental Resources in China." *Human Geography* 6(1): 102–125. https://doi.org/10.1177/194277861300600107

Colectivo Situaciones. 2007. "Something More on Research Militancy: Footnotes on Procedures and (In)Decisions." In *Constituent Imagination: Militant Investigations//Collective Theorization*, edited by S. Shukaitis, D. Graeber and E. Biddle, 73–93. Edinburgh: AK Press.

CREW. 2023. *Decommissioning Oil and Gas Infrastructure in California [Storymap]*. Santa Barbara: Center for Restorative Environmental Work.

Curley, Andrew. 2023. *Carbon Sovereignty: Coal, Development, and Energy Transition in the Navajo Nation*. Tucson: University of Arizona Press.

Dillon, Grace L. (Ed.). 2012. *Walking the Clouds: An Anthology of Indigenous Science Fiction*. Tucson: University of Arizona Press.

Dunlap, Alexander. 2019. *Renewing Destruction: Wind Energy Development, Conflict and Resistance in a Latin American Context*. London: Rowman and Littlefield.

Fanon, Frantz. 1963. *The Wretched of the Earth*. New York: Grove Press.

Frazier, Lessie Jo. 2007. *Salt in the Sand: Memory, Violence and the Nation-State in Chile, 1980 to the Present*. Durham, NC: Duke University Press.

Gilmore, Ruth Wilson. 2017. "Abolition Geography and the Problem of Innocence." In *Futures of Black Radicalism*, edited by G. T. Johnson and A. Lubin, 225–240. New York: Verso.

Gross, Lawrence William. 2014. *Anishinaabe Ways of Knowing and Being*. Farnham: Ashgate.

Grubert, Emily. 2020. "Fossil Electricity Retirement Deadlines for a Just Transition." *Science* 370(6521): 1171–1173. https://doi.org/10.1126/science.abe0375

Hall, Stuart. 1988. *The Hard Road to Renewal: Thatcherism and the Crisis of the Left*. London: Verso.

Healy, Noel and John Barry. 2017. "Politicizing Energy Justice and Energy System Transitions: Fossil Fuel Divestment and a 'Just Transition.'" *Energy Policy* 108: 451–459. https://doi.org/10.1016/j.enpol.2017.06.014

Heynen, Nik and Megan Ybarra. 2021. "On Abolition Ecologies and Making 'Freedom as a Place.'" *Antipode* 53(1): 21–35. https://doi.org/10.1111/anti.12666

Hopkins, R. 2008. *The Transition Handbook: From Oil Dependency to Local Resilience.* Totnes: Green Books.

Hopkins, Rob. 2019. "Transition Movement." In *Pluriverse: A Post-Development Dictionary*, edited by Ashish Kothari, Ariel Salleh, Arturo Escobar, Federico Demaria, and Alberto Acosta, 317–320. New Delhi: Tulika Books.

Howe, Cymene and Dominic Boyer. 2016. "Aeolian Extractivism and Community Wind in Southern Mexico." *Public Culture* 28(2/79): 215–235. https://doi.org/10.1215/08992363-3427427

Illich, Ivan. 1973. *Tools for Conviviality.* New York: Fontana.

The Invisible Committee. (Ed.). 2017. *Now.* South Pasadena: Semiotext(e).

Johnson, Gaye Theresa and Alex Lubin. (Eds.). 2017. *Futures of Black Radicalism.* New York: Verso.

Kallis, Giorgos, Susan Paulson, Giacomo D'Alisa and Federico Demaria. 2020. *The Case for Degrowth.* Cambridge: Polity Press.

Kenrick, Justin. 2011. "Scottish Land Reform and Indigenous Peoples' Rights: Self-Determination and Historical Reversibility." *Social Anthropology/Anthropologie Sociale* 19(2): 189–203. https://doi.org/10.1111/j.1469-8676.2011.00148.x

Krawec, Patty. 2022. *Becoming Kin: An Indigenous Call to Unforgetting the Past and Reimagining Our Future.* Minneapolis: Broadleaf Books.

Kropotkin, Petr Alekseevich. 1995. *The Conquest of Bread and Other Writings.* Edited by M. Shatz. Cambridge: Cambridge University Press.

Lennon, Myles. 2021. "Energy Transitions in a Time of Intersecting Precarities: From Reductive Environmentalism to Antiracist Praxis." *Energy Research & Social Science* 73: 101930. https://doi.org/10.1016/j.erss.2021.101930

Lohmann, Larry. 2015. "Questioning the Energy Transition." *Boletín ECOS* 33(dic. 2015-feb. 2016).

Luke, Nikki and Nik Heynen. 2020. "Community Solar as Energy Reparations: Abolishing Petro-Racial Capitalism in New Orleans." *American Quarterly* 72(3): 603–625. https://doi.org/10.1353/aq.2020.0037

Macey, Joshua and Jackson Salovaara. 2019. "Bankruptcy as Bailout: Coal Company Insolvency and the Erosion of Federal Law." *Stanford Law Review* 71: 879–962.

Maeckelbergh, Marianne. 2016. "From Digital Tools to Political Infrastructure." In *The SAGE Handbook of Resistance*, edited by D. Courpasson and S. P. Vallas, 280–297. Thousand Oaks, CA: Sage.

Partridge, Tristan. 2017a. "Resisting Ruination: Resource Sovereignties and Socioecological Struggles in Cotopaxi, Ecuador." *Journal of Political Ecology* 24: 763–776.

Partridge, Tristan. 2017b. "Unconventional Action and Community Control: Rerouting Dependencies Despite the Hydrocarbon Economy." In *ExtrACTION: Impacts, Engagements and Alternative Futures*, edited by K. Jalbert, A. Willow, D. Casagrande and S. Paladino, 198–210. New York: Routledge.

Partridge, Tristan. 2018. "The Commons as Organizing Infrastructure: Indigenous Collaborations and Post-Neoliberal Visions in Ecuador." In *The Right to Nature: Social Movements, Environmental Justice and Neoliberal Natures*, edited by E. Apostolopoulou and J. Cortes-Vazquez, 251–262. London: Routledge.

Partridge, Tristan. 2022. *Energy and Environmental Justice: Movements, Solidarities, and Critical Connections.* London: Palgrave Macmillan.

Partridge, Tristan, Javiera Barandiarán, Nick Triozzi and Vanessa Toni Valtierra. 2023. "Decommissioning: Another Critical Challenge for Energy Transitions." *Global Social Challenges Journal* 2(2): 188–202. https://doi.org/10.1332/NNBM7966

Partridge, Tristan, Merryn Thomas, Nick Pidgeon and Barbara Herr Harthorn. 2018. "Urgency in Energy Justice: Contestation and Time in Prospective Shale Extraction in the United States and United Kingdom." *Energy Research & Social Science* 42: 138–146. https://doi.org/10.1016/j.erss.2018.03.018

Powell, D. 2017. "Toward Transition? The Rise of Diné Energy Activism and New Directions in Environmental Justice." In *ExtrACTION: Impacts, Engagements and Alternative Futures*, edited by K. Jalbert, A. Willow, D. Casagrande and S. Paladino, 211–226. New York: Routledge.

Ranganathan, Malini and Eve Bratman. 2021. "From Urban Resilience to Abolitionist Climate Justice in Washington, DC." *Antipode* 53(1): 115–137. https://doi.org/10.1111/anti.12555

Shah, Tushaar, Neha Durga, Gyan Prakash Rai, Shilp Verma and Rahul Rathod. 2017. "Promoting Solar Power as a Remunerative Crop." *Economic & Political Weekly* 52(45): 14–19.

Temple, Christel N. 2010. "The Emergence of Sankofa Practice in the United States: A Modern History." *Journal of Black Studies* 41(1): 127–150. https://doi.org/10.1177/0021934709332464

Vizenor, Gerald Robert. (Ed.). 2008. *Survivance: Narratives of Native Presence*. Lincoln: University of Nebraska Press.

Whyte, Kyle Powys. 2020. "Too Late for Indigenous Climate Justice: Ecological and Relational Tipping Points." *Wiley Interdisciplinary Reviews: Climate Change* 11(1). https://doi.org/10.1002/wcc.603

Whyte, Kyle Powys. 2021. "Against Crisis Epistemology." In *Routledge Handbook of Critical Indigenous Studies*, edited by B. Hokowhitu, A. Moreton-Robinson, L. Tuhiwai-Smith, C. Andersen and S. Larkin, 52–64. New York: Routledge.

York, Richard and Shannon Elizabeth Bell. 2019. "Energy Transitions or Additions?" *Energy Research & Social Science* 51: 40–43. https://doi.org/10.1016/j.erss.2019.01.008

17

LISTENING AND BUILDING TRUST

Community-led conservation in Lincoln, Montana

Ryanne Pilgeram and Jordan Reeves

The problem: Conservation of open space is one tool for addressing climate change. How can we create more conservation on public lands when rural communities often rely on those lands for extraction to support their economy? Moreover, when many rural communities are initially hostile to conservation efforts, how can we make inroads?

The solution: The solution is a community-led approach to conservation in rural communities; it means listening to the community's concerns and finding solutions that can address them while championing conservation.

Where this has worked: Lincoln, Montana, currently has a community-created legislative proposal, "The Lincoln Prosperity Proposal," that would conserve 120,000 acres around their community. This proposal, in a place with many community members previously skeptical or outright opposed to conservation, was achieved because conservation groups, with The Wilderness Society (including Jordan Reeves) at the forefront, listened, built real trust, and worked collectively to address the issues the community raised. The Lincoln Prosperity Proposal has broad community support, and the community is pushing their senators (from both parties) to support this proposal in Congress. If it passes, it will create permanent Wilderness and other conservation designations on those 120,000 acres.

DOI: 10.4324/9781003437345-23

In *Upside Down: A Primer for the Looking-Glass World*, Eduardo Galeano suggests that the United Nations failed to include a fundamental right in the 1948 and 1976 *Declaration of Human Rights*: the right to dream. Though deeply critical of the economic and historical forces that created the contemporary condition in Latin America, Galeano finds hope in the right to dream, a right he sees as central to imagining a different future, to envision the world we want to build. Galeano asks, "How about we start to exercise the never-proclaimed right to dream? Suppose we rave a bit? Let's set our sights beyond the abominations of today and divine another possible world" (Galeano 1998:334).

Jordan, a coauthor of this piece, spent his formative years in Panama working with local fishermen on sea turtle conservation. His time there helped attune him to the connections between global economic forces and their impact on local communities – whether those communities are in Latin America or rural Montana – and introduced him to Galeano's work. So, it is unsurprising that the "right to dream" is central to our work in the Rural Communities Program at The Wilderness Society (TWS).

Today, political, social, economic, and environmental issues all demand our immediate attention, so taking the time to imagine a different future may seem indulgent. But the work of building a different future means imagining what that future might look like. Only then can we build it.

In our work on the Rural Communities Program, we are fortunate to find ourselves in a position that is part dreaming and part doing. We work in a nonprofit organization committed to confronting the climate crisis in several ways, but especially by focusing on public land conservation. We came to this work in different ways and in different seasons of our careers, but we share a steadfast belief that to build a new future we must center people and communities, most especially those that have been historically marginalized and excluded from the traditional conservation movement.

The Wilderness Society: dreaming and doing

TWS understands that land conservation is a key mechanism for addressing the climate crisis, and we are working to conserve 30% of all land in the United States by 2030. This aptly named 30 × 30 project is one for which TWS is well-suited. Founded in 1935 by an "organization of spirited people who will fight for the freedom and preservation of wilderness," by the 1950s TWS drafted the first legislation to conserve wilderness that formalized into The Wilderness Act of 1964. In the intervening years, TWS has led the charge to protect 111 million acres of public land (The Wilderness Society 2023).

While there are legitimate critiques of conservation and nonprofit work – that it can be top-down and funded and researched through a system that privileges a Western epistemology – our work helps showcase the ways that nonprofits like TWS can think more inclusively and expansively about how conservation can work in and with impacted communities as it addresses the pressing issues of our time.

When Ryanne started as the rural communities manager for TWS, it was a new position and program for TWS, organized around rural communities, economic transitions, and conservation, under TWS's Community-led Conservation Program. Ryanne came to TWS after a 13-year career in academia as a sociology professor. One area of her research focused on communities transitioning from extractive industries, such as logging and mining, to recreation-based economies. Her work on this topic made clear the tension between conservation, often blamed for a loss of jobs, and a community's economic future. She also came to TWS with deep roots in rural communities and a family whose past, present, and future are connected to extractive industries, making any attempt to resolve this tension between extraction and conservation more than just an academic exercise.

Jordan's career has centered on conservation for years. Before he took over as rural communities director, he was TWS's lead community organizer in Lincoln, Montana. Jordan was similarly shaped by growing up in a rural community that was embroiled in controversies about water and land use. His resolve to center people in his work was solidified by his time in Panama with the Peace Corps where he worked with communities reliant on fishing as their economic base. The work focused on preserving an undeveloped island for sea turtles and for traditional uses by local fishermen because the community saw the link between their survival and the turtles.

Building trust through community-led conservation efforts

A keystone piece of a future, more sustainable rural economy in the United States requires the conservation of public lands. Yet, most of our conservation efforts focus on and impact rural areas, which culturally and politically have often been wary of, if not hostile to, the conservation movement. Political polarization over several decades has fueled extreme anticonservation political trends in much of the rural United States, negatively impacting conservation outcomes in rural places where many important public lands remain. These anticonservation trends are buoyed by economic transitions that often lead rural communities to see conservation as antithetical to their economic well-being. Rural communities sometimes blame "environmentalists" for their community's transition from a booming to a bypassed community. Towns that had relied on clearing timber and the protections of unions now primarily offer jobs clearing tables for tourists. These economic changes are complex, and in fact often have more to do with diminishing union power, globalization, and changing technologies, but the "environmentalist" boogeyman often takes the blame (Pilgeram 2021).

Our job is to imagine a future that seems impossible in our political moment: a future in which rural, often politically conservative, people are engaged with and support conservation efforts. Moreover, it is a future that recognizes the diversity of rural places that include Sovereign Tribal Nations, Indigenous communities, immigrant communities, Black communities, as well as many communities that are interwoven with people of different backgrounds.

One of the places where TWS has been particularly successful at using a community-led conservation approach to conserving public lands is in Lincoln, Montana. Lincoln is an unincorporated community, with many residents who have experienced very divisive past legal battles over the extraction of timber and minerals such as gold and copper from their surrounding public lands, and thus have been highly skeptical of conservation efforts and their potentially detrimental impacts to the local economy. Yet, conservation work around Lincoln has been a long-term national goal since the community sits on a swath of public land that serves as a critical migration corridor for wildlife linking the Greater Yellowstone ecosystem with the Crown of the Continent ecosystem. This corridor provides an important habitat for plants and animals but is largely unprotected and still vulnerable to future potential resource extraction.

Yet, Lincoln has become a success story in the conservation world. Currently, based on goodwill and relationships, the work in Lincoln also yielded a high-profile conservation win, a legislative proposal for public lands protections put forth by the people in that community and their political leaders. The Lincoln Prosperity Proposal would create 120,000 new acres of Wilderness protections and conservation management areas around the Lincoln Valley. Importantly, the Prosperity Proposal also creates several other new types of land designations that were jointly crafted between TWS and community leaders, designations that protect wildlife habitat, improve public land access and outdoor recreation opportunities, reduce wildfire risk for the community, and provide long-term certainty for local livelihoods like ranching and sustainable timber harvest. While that proposal waits for congressional approval, the proposal is the basis for the US Forest Service Forest Plan Revision completed in 2021 which put the conservation protections largely into practice as the work toward a permanent designation via legislation continues.

This legislative proposal exists because of slow, intentional relationship building and careful listening in the community and because TWS took a different approach to conservation in Lincoln. One of the key contributions from Jordan was a commitment to work on issues beyond just what was happening *on* public lands and to recognize the connection between the land and other community issues and to commit time and resources (and heart) to work on those issues in addition to longer-term big picture land management efforts that led to a legislative proposal.

Thus, in addition to conservation, TWS's work in Lincoln sought to help the community engage with their public lands while creating opportunities for them to imagine their community's future. Importantly, the work has included tangible strategies such as partnering on grants to build recreational trails, finding funding and partner organizations to help the community create an economic development plan, and even networking to convince TWS funders to financially support the locally led development of a landscape-inspired sculpture park in Lincoln.

That park, named Blackfoot Pathways: Sculpture in the Wild, now draws 50,000 visitors a year. One of the central features is a former "burner," a structure that was used to process timber by-products during the heyday of Lincoln's extractive past.

The park managed to find an intersection between the area's historic timber industry and the contemporary art community – not an easy needle to thread.

In another example, an open-space grant secured in partnership with the community and led by a local land trust finally gave the children of Lincoln access to the legendary Blackfoot River that flows right through the middle of town. Previously, there was no public access to the river, and instead, kids had to scramble down an embankment next to a highway bridge.

These conservation victories were achieved not simply by TWS as a national nonprofit pushing for change but by creating collaborative relationships with local groups and local communities that had historically been antagonistic to conservation efforts, based on their past experiences. By building relationships, listening to community members, and rethinking our own visions of conservation, we were able to find common ground concerning conservation's place in the community's future.

TWS's work in Lincoln helped a community well-known for skepticism and conflict with environmentalists begin to trust a conservation organization enough that Lincoln community members continue to ask that TWS be a partner. They keep saving TWS a seat.

The Wilderness Society and Lincoln, Montana, as a road map

These successes in Lincoln serve as a road map for how to do conservation work in other rural communities. The scope of the climate crisis and related issues are enormous, but there are models already in place to address a crucial part of these issues. Those models entail working in the community, and they require deep and long-term investments in relationships.

Even in communities that are not hostile to the conservation movement, environmentalists and conservation groups can often be seen as flaky, out of touch, and unreliable. TWS's work in Lincoln sought to challenge those stereotypes by demonstrating authentic, long-term engagement with local peoples and communities. We wanted to be a committed partner in the long-term health and well-being of the community and be recognized as such by local people.

This yielded a variety of concrete strategies. For example, TWS hired community organizers and other positions from within the community. One of Ryanne's first conversations, when she started at TWS, was with an organizer from Lincoln who was hired as a consultant to support the Lincoln Prosperity Proposal and has played a critical role in that success. The success in Lincoln is, in part, a testament to her skills at community organizing and her ability to listen to the community. She had been working at a local restaurant and was privy to all sorts of conversations about people's concerns. One of the things she heard was from hunters who were worried that it was getting more and more difficult to find game on public lands because those lands were not being well maintained. The wildlife was heading to private land to graze. Our political landscape has led us to believe, we think to our detriment, that there are these vast irreconcilable differences between conservative-leaning rural communities and conservation goals, but what she heard (and, more importantly, what she understood) was that people in rural communities are intrinsically invested in the health of our public lands. People in rural communities have deep ties with the land as well as great appreciation for the clean air and water, wildlife, and wide-open spaces that define a rural lifestyle. In fact, they have often made a conscious choice to stay in rural communities despite greater economic opportunities, better access to healthcare, and many more services and conveniences available in suburban or urban areas. These kinds of insights were vital to forging relationships with the town.

As this suggests, TWS has also prioritized community goals that were not necessarily a part of our conservation efforts but helped the overall health of the community. The sculpture park in Lincoln, public access to the river, completion of an economic development plan, and the construction of recreational trails on public lands made clear that our goals are not simply conservation but supporting the community and its engagement with the surrounding lands.

Beyond the work in Lincoln, TWS has taken a similar people-first approach to conservation in communities in many other places, further connecting dots between community well-being and conservation in states as different and far-flung from one another as Alaska, New Mexico, Idaho, Arizona, California, Colorado, Washington, Maine, and North Carolina. These kinds of local investments demonstrate a long-term commitment that builds essential trust and opens the door to conversations about conservation, and are a central pillar of the TWS Community-led Conservation Team within which our Rural Communities Program is nested. For example, we have hired local grant writers in many rural communities to advance community priorities on a wide range of issues that stretch beyond a traditional conservation focus. TWS has supported community economic development and diversification plans, devoted staff time to launching local farmers' markets, assisted with grants to expand local agricultural production and food systems, and lent expertise to study opportunities to expand local housing options. Seeing a major need to make outdoor recreational opportunities more equitable in rural communities, and a clear nexus between recreation and our organizational focus on public land management, we have also devoted staff time and resources toward securing private land for new community parks, designing and building community trail systems, expanding local efforts to connect youth to the outdoors, and helping communities market their outdoor recreational assets to potential visitors and future residents.

TWS's efforts to advance community priorities alongside conservation gains have also expanded beyond just local action in the rural communities with whom we partner. Utilizing our organizational expertise at creating and advancing national policy, we have helped to shape and pass new policies that support rural communities across the United States in mitigating past environmental harms caused by extractive industries and restore the ecological health of lands and waters impacted by extraction, targeting millions of dollars in federal investment to rural communities and creating thousands of rural jobs. We have also used advanced policy to resolve challenges facing many rural communities and create new, innovative management policies to protect cultural values important to rural community members. For example, TWS successfully advocated for a new policy requiring a local Forest Service District in Montana to address land management needs by comprehensively planning to combat noxious weeds, a key issue brought forward by community members and agricultural producers. We have also advocated for new federal land designations that would protect critical wildlife habitat values while also preserving long-standing agricultural practices important to communities, such as grazing, prescribed fire, and forest restoration. Finally, and more recently, we have advocated alongside rural community partners for land designations recognizing and preserving cultural heritage.

TWS's efforts in Lincoln and elsewhere worked because we listened to local communities, invested in local communities, and distinguished ourselves as empathetic, trustworthy, and solution oriented. This model can serve as a theory of change for conservation work that better connects rural communities and the national conservation movement. Ryanne is originally from this area of Montana, and to say that building collaborative conservation in Lincoln seemed impossible on paper is an understatement. It feels especially daunting because building long-term relationships in rural communities can't be described in an easy formula (philanthropy dollars + earned media + community organizer hours = success), but we still know it works and that it can be duplicated. Using Lincoln as our example, our theory of change starts with an individual or organization, who serves as a change agent to support innovative projects that are community driven but that would likely not be otherwise attempted due to a lack of capacity or potential concerns that the idea was too disruptive or innovative in nature.

The first step may be to demonstrate success on small-scale projects at the community level, for example, the river park in Lincoln. Connecting that success with the values of community cohesion and conservation can effectively demonstrate trust that conservation-focused projects can deliver holistically for the community.

Building trusted partnerships by delivering on community ideas allows leaders to see both the linkage between the conservation of their public lands and their community's health and allows community leaders with political and social capital to more actively connect those successes to advocate for conservation-based solutions in their communities, solutions that may have previously seemed untenable.

Encouraged by their successes, community leaders can voice conservation-based policy and legislative solutions at the county, state, and federal levels with less personal risk, and with the backing of key constituents, city, county, state, and national leaders can enact more conservation-focused policies.

While we believe these strategies help attain specific goals such as conserving land and abating climate change, these strategies also aspire to give people agency in their own backyards. At the core of what we're describing is a process of how communities can move from extractive economies to economies that allow communities to treasure, steward, and connect to their landscapes for the long term. Connecting to place is often a primary indicator that one will be moved to action to address climate change, whether by supporting the conservation of public lands or in other aspects of life. The Alaska Just Transition Collective stresses, "it is essential that we orient ourselves to a deep understanding of the language we use to describe the future we want to build." One of the tasks before us, then, is to imagine not only the work we will do but the language we will use to describe it.

The Alaska Just Transition Collective is one of the first groups to adopt this language in their "Regenerative Economies" report. They emphasize that the root of the word economy is "management of home" and argue that the great challenge of our time is how we will "build and nurture sustainable communities, social, cultural and physical environments which satisfy our needs and aspirations without diminishing opportunities for future generations" (2022:7).

TWS's work in Lincoln has not only yielded a proposal for land conservation, it has helped shift community perceptions about public lands. Many people in Lincoln tended to characterize nearby public lands as a burden, poorly managed, untapped for their logging and mining potential, and posing a dramatic wildfire risk to local community members. We are supporting the community in a shift to a more hopeful vision for their backyard, their home, that celebrates public land as an asset that supports a sustainable, thriving future for the community.

We have also been a part of supporting the community to achieve a greater sense of autonomy in defining and shaping their relationship with the lands around them. A key insight from one of the local community organizers was that the town had no idea that they had a say in what public land management looked like in their area. TWS's community organizing in Lincoln helped the community to better understand complex forest management policies, legislative processes, and rural development grant programs. These are tools that had always been available to the community but whose complexity and bureaucracy were daunting. Of course, these tools are not just daunting for people in Lincoln but for many, even most communities who are not versed in navigating them and whose leaders are often juggling multiple volunteer roles to meet community needs. By assisting the community in navigating this complexity, TWS opened the door for Lincoln to shape the future of forest management and conservation in their area, helping the community to write new narratives about their home and their public lands.

We are at a transformative moment nationally with an unprecedented amount of federal investment up for grabs intended to support rural communities in addressing climate change. TWS and other nongovernmental organizations (NGOs) are also investing more resources than ever in community-led conservation, Indigenous-led conservation, and Just Transition work. The Infrastructure Investment and Jobs Act of 2021 funnels an unprecedented amount of funds into renewable energy, climate adaptation, and rural development, and there is tremendous work happening on the ground by many community leaders to reimagine our collective future.

The challenge we face is how to take these opportunities and move from short-term, single-issue conservation wins toward a long-term transformation of the sociopolitical landscape around conservation and community well-being. We live in an economy and political system that measures success in short-term markers – annual return-on-investment to shareholders or the number of candidates elected from a political party – but success in this work requires deliberate investment in relationships, and relationships built on trust take time.

As we invest deeply in our rural communities, we must be guided by the principles of collaboration and listening. The communities where TWS is working on the surface might seem quite different. In Lincoln, Montana, they are reeling from the loss to the timber and hard rock mining economy. In New Mexico, the growth of oil and gas extraction shapes their relationship with public lands. In Appalachia, the relationship to extraction is magnified as they face pressure from the loss of coal, the growth of natural gas, and continued pressure from logging. In tribal nations and Indigenous communities, these impacts are amplified against the work to heal from the violence of settler colonialism and assert their sovereignty and treaty rights. Yet, what connects these examples is that it is the people themselves who should decide what the future of their communities will look like.

While we firmly believe those are questions for communities to answer, there is no doubt that the answers must be rooted in a future that is at its core regenerative, a term that connotes a certain kind of energy, because regenerative powers cause something to heal or become active again after it has been damaged.

All of these challenges stem from the fact that these communities were designed by and for extractive industries, but those industries were not designed to center the people who worked in them, nor the landscapes where they were drilling and logging. The fact that communities have survived in spite of the booms and busts of extraction is a hopeful sign of their resiliency and their regenerative potential. But across rural America, we should remember that most communities were not built to last – they were built to create an economic boom for the industry while there were raw materials to extract and were organized in many ways with little concern for what happens next.

Our challenge today is to envision a sustainable future for those communities, a vision that can only be crafted by working closely with people in those areas. Working at the forefront of conservation – and thinking expansively about the relationship between communities, extractive industries, and public lands – means listening to communities and supporting them as they begin the deep work of imagining a different future.

Conclusion

We understand that the next decade very well may include greater political polarization, starker socioeconomic divisions between urban and rural communities, and an extreme anticonservation agenda among many of our political leaders. But we are optimistic that this work can transcend traditional partisan political divides because we have seen it happen already. We are serious when we say we think working in, with, and for the community can be transformative.

We sometimes need to remind ourselves that despite the challenging political headwinds for advancing conservation goals and protections that conservation has faced in recent years, we have been very successful when we have invested in long-term, authentic partnerships rooted in local landscapes and with local people. Those successes endured political sea change. In fact, they have largely avoided the attacks and rollbacks leveled at many shorter-term national gains over the last decade.

Our work has shown that when we invest more in deep partnerships that bridge concerns about conservation, livelihoods, and local economies, then conservation and communities win. We also know that one of the strengths of an established NGO like TWS is connecting our grounded local work in communities to policy and legislative solutions in Washington, DC.

None of us have all the answers to the complex problems we are seeking to solve, and the history of nonprofits, and conservation nonprofits, is fraught. There are so many different futures each of us must imagine, contributions we can make, and expertise we can offer, but centering relationship to build power with communities is an important path forward.

References

Alaska Just Transition Collective. 2022. *Regenerative Economies: A Guide to a Thriving Alaska*. https://issuu.com/justtransitionak/docs/regenerative_economies_a_guide_to_a_thriving_alas

Galeano, Eduardo. 1998. *Upside Down: A Primer for the Looking-Glass World*. Translated by M. Fried. New York: Picador.

Our History. 2023. The Wilderness Society. www.wilderness.org/about-us/our-team/our-history#:~:text=The%20Wilderness%20Society%20was%20founded,being%20built%20on%20the%20Dinosaur

Pilgeram, Ryanne. 2021. *Pushed out: Contested Development and Rural Gentrification in the US West*. Seattle: University of Washington Press.

18

CATTLE GRAZING AND CLIMATE CHANGE ADAPTATION

Local environmental knowledge and public lands management in the US West

Chloe B. Wardropper and Nicolas T. Bergmann

The problem: If climate change solutions formulated by scientists and policymakers fail to account for local environmental knowledge and people's cultural connections to landscapes, they may fail to acknowledge societal trade-offs from proposed solutions, fail to get local buy-in, and miss important traditional approaches to dealing with climatic variability.

The solution: Acknowledge context and integrate local environmental knowledge into mainstream analyses and recommendations for mitigation and adaptation.

Where this has worked: Some cattle ranchers in the western United States practice vertical transhumance and use their local environmental knowledge to adapt to climate change in multiple ways, particularly by moving livestock across a large mountainous landscape according to seasonal and annual variability.

"It's already hot, too damn hot. It's only late June," thinks Ed as he hops down from his white Chevy pickup. He squints into the morning sunlight but can't locate his cows. He figured they'd be down near the stream by the fish exclosure, but maybe they've moved over into Gold Creek on the other side of the Forest Service allotment. Ed jumps back in his truck and rattles off down the washboard road scanning the forest edge. The radio crackles in the background – something to do with a "heat dome" that's descended across the region. Ed doesn't really want to think about what the grass will look like in August, but he knows he'll have to. Hay is going to be a tough deal this year.

DOI: 10.4324/9781003437345-24

Ed's driving slowly. He's got a 1,000-gallon water tank strapped on the trailer. Water is often the limiting factor for the operation, and Ed's proud of the mobile solution he's concocted. Sometimes he wishes the cows were more like sheep, at least when it came to needing less water. But they're not, and he knows he needs to better utilize the available forage this year and distribute the cattle across the range. Ed finds the cows nestled in a shady area farther down the stream. He doesn't want to stress them too much but gets them moving. He knows they'll appreciate the fresh forbs this evening, and he doesn't want the riparian areas to get hit harder from grazing than they have to.

It's hot here but not as bad as on the Snake River Plain; the elevation moderates the heat. It might not be as good as air conditioning, but it helps. Ed knows that the key to navigating a bad drought is adaptability. In years like this, the cows like to move up the mountain early. There's already open grass up there, but that needs to be saved for next month. It'll take more management this year to hold them back, but that's okay. Ed would rather avoid office duties anyway. After moving the cows and unloading the water into a mobile stock tank, he heads back to reload. He figures it's time to call in some favors and see if his hay suppliers can cut him a deal. Ed knows that close friendships may be the best drought mitigation strategy. He's worked hard at cultivating those over the years.

Ranching and climate change in the western United States

Ed is one of a declining number of ranchers in the US Intermountain West grazing cattle on public lands. This region has seen rapid social and landscape change over the past few decades, with the consolidation of family ranches and move toward an amenity-based economy (Swette and Lambin 2021). Although there are fewer cattle operations on western rangelands today than in the past, the US cattle industry still produces more beef than any other country in the world (Greenwood 2021). People outside and within the industry are concerned about its impact on climate change, and vice versa. Beef production generates significant greenhouse gas emissions (Eisen and Brown 2022; Gerber et al. 2015). At the same time, climate change impacts like extreme temperatures, larger and more frequent fires, lack of water, and increased disease prevalence threaten both the cattle industry and the health of rangelands, particularly in the western United States (Chambers and Pellant 2008; Godde et al. 2020; Reeves, Bagne, and Tanaka 2017).

US rangelands can be conceptualized as underutilized spaces for capital investment, with livestock grazing a low-profit "placeholder" for land use until more lucrative opportunities come along (such as residential development or energy production) (Sayre 2023b). Increasingly, rangelands are valued for providing other ecosystem and social services like biodiversity conservation and recreation (Maher et al. 2021; Sala et al. 2017). Consequently, environmental groups have led litigation against the Bureau of Land Management and US Forest Service – the two agencies that manage most public rangelands – to try to reduce the impacts of land uses like ranching on water quality and wildlife habitat (Martin 2021; Nie and Metcalf 2016).

Furthermore, climate activists and some researchers call for a move away from meat-based diets (Eisen and Brown 2022), while the 2022 Inflation Reduction Act prioritizes renewable energy production in the United States, including on rangelands. Yet livestock grazing is a less-intensive land-use option than development, and grazing is a beloved rural livelihood in the western United States (Brunson and Huntsinger 2008; Swette and Lambin 2021). These competing visions of sustainability typify land-use conflict in the United States and are only intensifying as climate change progresses.

We, two social scientists with Western scientific training, are concerned with questions of livelihoods and sustainable production in US agriculture. We want to know: is there a future for cattle grazing on US rangelands under climate change? What can ranchers' experiences teach others about responding to climate variability? And how can mainstream climate change activists and researchers, whose views are often different from those of people living and working in rural communities, better include other ways of seeing and addressing climate change?

In this chapter, we describe the urgency of engaging local environmental knowledge, especially understanding ranchers' views and experiences related to climate change in conservation efforts. We tell two additional ranchers' stories to illustrate our points. Like Ed's story, these are composites. We give the ranchers fictional names and draw from multiple individuals' stories. All three stories are derived from 44 interviews and ethnographic research we conducted in 2022 with ranchers, rangeland managers, and wildlife managers in the US Intermountain West region of Idaho, Oregon, and Washington.

Local environmental knowledge and context

There is a growing subfield across several social science disciplines that documents how different ways of knowing the landscape, or local environmental knowledge (LEK), affects perceptions of and action on climate change. This body of work investigates how people's views on changing weather are tied closely to the material conditions of their lives. LEK is derived from practice-based knowledge of how a person manages their livelihood, which varies depending on the livestock or crops they tend, technologies they use, cultures they belong to, and ecosystems where they work (Burnham, Ma, and Zhang 2016; Ingold 2000; Ingold and Kurttila 2000).

The field of LEK can help readers of this book understand how US ranchers think about and adapt to climate change. We acknowledge that this field has typically studied local knowledge in the context of Indigenous people's traditional knowledge (Fernández-Giménez et al. 2021), whereas US ranchers primarily belong to settler-colonial groups (Huntsinger et al. 2010). Yet LEK literature helps explain these ranchers' intimate experiential- and place-based knowledge. Furthermore, many scholars working within the field wish to add nuance to simple dichotomies between "local" and "scientific" knowledge (Horowitz 2015). US ranchers often draw on multiple forms of knowledge, blurring the binary distinction between Western scientific and LEK. Going to schools that teach the scientific method, and interacting with or working for government agencies that aim to implement "evidence-based" programs, most US ranchers are well-versed in Western scientific culture. But many of the ranchers we interviewed think about their ranching operation holistically and draw on lived experiences, neither of which are easily reduced to hypothesis-driven scientific research.

In the US West, rangelands are split between private and public ownership with many ranches operating across multiple types of parcels. Grazing on US federal lands, formalized in the 1930s through a system of leased allotments, is a governance structure that helps sustain ranch families through affordable grazing leases that complement private rangeland ownership (Sayre 2023a). Some ranchers move cattle from lower to higher elevations in a seasonal pattern to take advantage of available moisture and the timing of plant growth, a practice called "vertical transhumance" (Huntsinger et al. 2010). Because a large amount of higher elevation rangeland in the US West is owned by the federal government, public policy significantly influences the viability of this type of adaptive ranching system (Huntsinger et al. 2010; Swette and Lambin 2021).

Bart's story

It's an unseasonably cool June morning but Bart is on horseback and smiling. He waited about a month to take advantage of spring growth on his Forest Service allotment in Hells Canyon Country – rugged lands straddling northwestern Idaho and northeastern Oregon. Now his 250 mother cows and their calves are finally out grazing. The sun is just coming over the tops of the trees as he rounds a bend in the trail. We say "trail" but it's really a decommissioned logging road that's slowly starting to look more rugged every year. Bart pauses at a viewpoint. Several thousand feet below, he sees the river snaking its way along the canyon bottom. He can't quite make out the ranch house, but it's down there. After readjusting in the saddle, Bart glances up to see the last snow fields nestled into the lower portion of an alpine meadow at the base of the mountain several thousand feet above. The winter was hard, and it's nice to be in the uplands again. Being on horseback moving cows is his passion and helps make up for all the frustrations that come along with being a cattle rancher these days.

Bart is a fourth-generation rancher. Raised down in the valley, he was a high school football captain and went off to play for 4 years in college. Along the way Bart earned a degree in plant sciences and thought he might work in a lab as a plant breeder to make some real money. But ranching is in Bart's blood, and he couldn't pass up the opportunity to come join Dad after his football days were over. Bart sees the landscape through the eyes of a stockman and wants the range to flourish for future generations. Bart is passionate about perpetuating drought-tolerant native perennial bunchgrasses[1] – especially Idaho fescue – that his mother cows turn into milk for calves. Bart learned about plants from his coursework in college, but most of his knowledge is from the 30-plus seasons he has spent observing how plants respond to livestock grazing across a range of environmental conditions. Bart understands how a diversity of plant species not only benefits his cattle and the land but also a multitude of other living organisms. Drought has gripped the region in recent years, but Bart feels fortunate that the care his family has put into managing the rangeland is paying dividends. He also knows that putting a ranch together in such rugged country has inherent biogeographical advantages.

Bart's ranch doesn't experience drought in the same way that many others do, because the ranch practices a form of vertical transhumance. In this system, winter pasturing and supplemental feeding occur on low-elevation private land, while during the spring, summer, and fall months, his operation relies on a combination of federal and state public lands to graze an elevation gradient of more than 5,000 vertical feet. These public lands – especially the Forest Service allotment – provide ecosystem diversity and allow for ranches to turn the limited moisture of the semi-arid US West into pounds of grass-fed beef. Furthermore, ranching keeps the land open for wildlife habitat, including sage grouse and Rocky Mountain elk. Bart is proud of this model of sustainable food production and land stewardship and often laments the lack of value that many urban folks attribute to public lands grazing. Glancing backward over his shoulder, Bart sees the edge of the fire that happened two seasons ago. He blames poor federal management for the severity of the damage and worries that a warming climate and burdensome regulations will exacerbate the problem and limit his ability to sustainably manage the land.

A wildfire in 2020 burned almost 55,000 acres and almost took out half the herd. After the immediate stress of the fire passed, Bart feared for the financial viability of the ranch. In a different area that burned previously, the Forest Service shut down grazing for 2 years. Thankfully, Bart has a great relationship with the local range con[2] who understood not only the intricacies of grazing in the aftermath of fire but also the importance of forage availability to keep ranches operating. Working collaboratively, they came up with an adaptive management strategy to allow grazing in portions of the burn that had received minimal impact while staying out of areas that were torched. That plan relied heavily on Bart's skill to manage his cattle as well as a deep understanding of how plants respond to stressors. It hadn't been an easy process, but he was satisfied with the outcome; Bart only had to destock 20 mother cows and thinks the grazed areas are recovering well.

Despite the good work of the local range con, Bart couldn't help but throw up his hands in frustration at certain environmental policies that he thinks exacerbated the 2020 fire. In Bart's eyes, the Forest Service's departure from more active forest and range management in the name of recreation and biodiversity, including increasing restrictions on public lands grazing to provide protection to certain endangered species and improve water quality, has produced a landscape more prone to catastrophic fires that in turn threatens biodiversity. When combined with warming temperatures and decades of fire suppression, public lands ranchers know they provide value keeping fuels in check through targeted grazing but remain frustrated at policies that make it difficult, particularly the Endangered Species Act (ESA) and the National Environmental Policy Act (NEPA).[3] Restrictions mean that Bart is often unable to graze pastures at the optimal time and cannot adjust his grazing plan to best account for seasonal or climatic variability. For example, warm early season growing conditions the previous spring meant that higher-elevation pasture was ready for grazing sooner than his forest permit allowed. Instead of being able to graze when the range was ready, he had to wait several weeks to turn the cattle out. Consequently, he missed an important grazing window and put higher grazing pressure on other parts of the ranch. Furthermore, the lack of Forest Service capacity to navigate ESA or NEPA regulations often means that adaptive management strategies, like additional water infrastructure development to deal with drought, are in limbo.

Bart understands that grazing on public lands is one of many uses but feels that the government and the American public need to support the adaptive capacity of ranchers practicing more sustainable ways of food production. In Bart's eyes, the environmental knowledge obtained through generational experience and careful observation far outweighs any recent scientific modeling derived from high-powered satellite imagery that often informs federal land management actions.

Carla's story

It's Halloween and snow is falling lightly in Oregon's Blue Mountains. Carla is flying on the side-by-side trying to locate part of the herd. It has been a good fall following a hot summer. Early September rains provided some extra growth and enabled the cows to stay on the allotment through the end of October. Ahead of her she sees the fresh snowline – bright white against an autumn palate of orange and brown. Carla hops off at the entrance to the large meadow. She finds the stake and snaps a quick photo. A feeling of pride overwhelms her. The meadow looks good. She can't spot any sign of medusahead, and there's not much cheatgrass either. The grazing plan is working. Native species are thriving, and forest encroachment has stalled. With another quick scan, Carla estimates utilization at around a quarter.[4] The meadow could handle more cattle. In fact, the entire allotment needs more grazing.

Carla just celebrated her 45th birthday. Her older brothers both left the ranch to become doctors, but they come out twice a year to help with calving and hauling cows home in the fall. Carla had planned to become a veterinarian, but then Dad's accident happened, and things changed fast. She'd always loved the animals but wasn't sure if she was cut out for daily ranch life. It turns out she was. She chuckles and thinks about how she probably likes plants even more than animals these days. And that's saying something. Carla had always gravitated to helping with the livestock. She knew what a healthy cow looked like, but it wasn't until her partner gave her a copy of Savory's book that she really began to think about the environmental benefits of grazing.[5]

Dad always cared for their ground but never really looked to expand the business and take on more cattle. Carla was doing more to sustainably grow the ranching operation: making the ground more productive and securing additional pasture. Dad took a conservative grazing approach and always kept a low density of animals, which made some sense and still does. Earlier cattlemen overgrazed, and logging made a mess of the forested land her paternal grandparents bought in the 1950s. Lower grazing pressure helped the land, but there were still plenty of weedy infestations and marginal pasture when Carla took over. Although she'd acquired a lot of knowledge through years of observation and partnering with the university on a variety of research projects, she hadn't put it all together until she read Savory. A holistic approach to ranching made so much sense. What other approach connected the realities of ranch economics with the mineral cycle?

Carla's mantra is that the ground never lies. If you want to see how your grazing practices are affecting the range, trace it through time and use your eyes. She's made loads of mistakes, but that's what she loves about holistic management. Mistakes are built into the process, but that's how you learn. The variability of soils, plant communities, and moisture regimes on her ground meant that high-intensity, short-duration grazing didn't work in every spot, but generally the positives outweighed the negatives. Carla laughs when her friends tell her she takes grazing management too seriously. She does spend more time moving cattle than anyone she knows, but in her eyes that's what ranchers are supposed to do. Moisture is often the limiting factor to grazing management in the Intermountain West. She's not afraid to haul water to get optimal grazing pressure where it's needed. She's still working to figure out a better system on the forest allotment. She's made progress, but it's a struggle to bunch graze cattle and find proper water in remote forested uplands.

Carla likes that there's room for improvement. She knows she's been lucky to escape fire and hopes that her permit can be amended to add more AUMs.[6] The federal bureaucracy moves slowly and can hamstring those with a more holistic approach to grazing management. Better grazing management has led to more forage growth and more carbon sequestration. The land can now handle more animals, even in drought. It would be a shame to let all that progress go up in flames. Sometimes she wished she could just get rid of the bureaucratic headache that comes with the public allotment but is thankful for the ability to capture moisture across an elevation gradient.

Carla is worried about climate change. Deeply worried. There's a lot of political grandstanding about how to move forward. She believes most of the proposed solutions will cause more harm, especially those that favor converting public and private rangeland to more intensive uses. Carla hates the thought of losing wildlife habitat to sprawl and subdivisions, and converting working lands to solar farms in the name of improving the state's carbon footprint really drives her bananas. About once a month she's approached by one of these companies, but she'll never sell out. Her ranch sells and markets cattle through the regional beef co-op. It's nice to escape the tentacles of big beef where she can. Plus the higher prices she gets for her product pay for more expensive grazing practices like keeping range riders out with the cattle. She just wishes people buying grass-fed beef in the big city would make the connection between the product and good land management.

It's getting late, and Carla is perplexed. She's covered most of allotment's territory but still hasn't located the last group of cattle. Just then her radio crackles. Sean, her eldest brother, found them down near Elk Creek, and they're all bunched up. It sounds like the dogs are spooking them. Carla calmly tells Sean to put the dogs in the truck and floors it. She's able to drop 2,500 feet and cover the 5 miles in just under 15 minutes. It takes a long time to round the cows and calves into the trailers, but eventually it's done. The ear tags and scanner make it easy to count, and they've got everyone. It's about 100 miles to home, and they practice transhumance with diesel by necessity these days. She hopes some more allotments will open back up closer to home but won't hold her breath waiting for the feds to move on that.

Using local knowledge to mitigate and adapt to climate change

Like Ed, Carla and Bart are deeply embedded in their ranching systems. Their operations are centered around the health of perennial bunchgrass ecosystems, relying on seasonal transhumance across elevation gradients, and their investment in ranching is due in large part to family legacies and cultural connections to the land. Ed, Bart, and Carla's stories of managing public and private lands in the western United States bring up many of the ways ranchers are affected by and adapt to climate change. Climate change contributes to more frequent and severe fires like the one Bart battled. Climate change creates conditions for invasive plants like medusahead to spread, crowding out native grasses that provide the best nutrients for cattle; Carla uses concentrated "bunch" grazing and other strategies to reduce invasive plants. Climate change is exacerbating seasonal variability like spring green-up and the timing of plant senescence; Ed, Carla, and Bart's transhumance systems allow them to move cattle to pastures as they produce the most vibrant and nutritious fodder (though permits might constrain their choices). Climate change is increasing the frequency and severity of drought conditions, forcing some ranches to sell cattle before they reach their optimal size if there is not enough food or water available; Carla's membership in a regional beef co-op means her meat sells for higher prices so she can keep her herd smaller if needed.

Ranchers like Ed, Carla, and Bart have come up with many adaptations to climate change, though they might not label them as such. Saliman et al. (2022) argue that descriptions of adaptations on US ranches should focus on the processes ranchers already employ to respond to change. Often, short-term changes become long-term solutions to new climate normals. Short-term changes might include finding new sources of hay, selling down the herd, trucking in water, and switching up pastures in a bad drought year. These changes might become permanent in the form of new stable herd sizes, different pasture rotation systems, or stream restoration projects.

Ranchers have valuable lessons to teach others about using an entire system to manage seasonal and annual variability. Practicing transhumance helps reduce drought stress by taking advantage of seasonal changes across elevations to sustain a ranching operation through seasonal and interannual change. Ranchers are also adept at combining insights from Western science with local knowledge. This skill is highlighted in Carla's story describing her use of a holistic grazing approach, coupling historic and predicted climate information with detailed place-based knowledge to make plans for rotational grazing. While there is debate in the ranching community about the promises and pitfalls of holistic management (Gosnell, Grimm, and Goldstein 2020), the approach reflects the type of integrative thinking characteristic of livestock producers with semi-extensive operations. Bart's approach to fire recovery is another example of adaptation in a large landscape system – assessing burn severity across allotments to target areas that can handle grazing pressure. Targeted grazing post-fire has also been used as a strategy to battle another climate change impact: the spread of invasive plants, which can take root in bare soils after a fire (Porensky et al. 2018). Other adaptations represent hard work coupled with landscape knowledge, like Ed's use of mobile stock tanks to supplement water and reduce the grazing burden on riparian areas. Finally, some ranchers have taken a collective action approach to creating a more sustainable market for their beef by creating co-ops like Country Natural Beef and Desert Mountain Grass Fed Beef. These co-ops counteract some of the consolidation in the meat industry that can lead to undesirable social and environmental outcomes.

Public lands ranchers argue that transhumance is a climate-resilient form of livestock production. It preserves cultural values, like supporting rural communities and ecological values, managing invasive species, reducing fire risk, and preventing conversion of open lands to development. Supporting those values into the future will require greater acknowledgment of ranchers' LEK that can help adapt to climate change–related problems. Support could come in the form of increased flexibility within public lands grazing leases so that ranchers can move cattle in response to fire, drought, and water changes. National-level policies addressing climate change concerns on rangelands would benefit from grappling with the trade-offs that come from different land uses and taking local people's knowledge and management approaches seriously.

While there are many opportunities to mitigate and adapt to climate change, there are no easy solutions. Making large-landscape shifts toward renewable energy or moving away from meat production have implications beyond climate mitigation that we must grapple with as a society. Furthermore, adaptations to climate change impacts are context dependent. This means there is no one-size-fits-all approach to solving the climate crisis, and almost all solutions contain complex trade-offs. As a society, we would do well to craft careful climate policy that accentuates landscape resiliency. Paying careful attention to peoples' local environmental knowledge is a great place to start.

Acknowledgments

The research discussed in this chapter was supported by grant # 2317537 from the National Science Foundation. We would like to thank the ranchers and land managers who took part in our study.

Notes

1 Perennial bunchgrasses anchor the native composition of many rangelands in the US West along with perennial forbs and shrubs. These plant species have extensive root systems that are key for maintaining soil quality and a healthy plant community and limiting the growth and spread of annual invasive grasses such as cheatgrass and medusahead.
2 "Range con" is short for range conservationist. These agency employees work with rancher lessees to manage US Forest Service (and Bureau of Land Management) rangeland allotments.
3 The Endangered Species Act regulates land-use actions that could harm officially listed threatened and endangered species. When an allotment contains a listed species, grazing can be restricted at certain times and locations to reduce impact. The National Environmental Policy Act requires environmental assessments and public engagement for federal projects and on federal lands. This can make it difficult to justify certain actions on public lands or make agencies managing grazing permits vulnerable to lawsuits.
4 "Percent utilization" is a measurement rangeland managers use to monitor and evaluate how much forage has been grazed by animals.
5 Allan Savory is a former Zimbabwean politician and wildlife biologist who created the framework of holistic management to improve grazing management.
6 An AUM or Animal Unit Month is a standardized measurement of rangeland carrying capacity used to estimate the forage requirements of a cow (or other livestock) and the number of cattle a given landscape can sustain.

References

Brunson, Mark W. and Lynn Huntsinger. 2008. "Ranching as a Conservation Strategy: Can Old Ranchers Save the New West?" *Rangeland Ecology & Management* 61(2): 137–147. https://doi.org/10.2111/07-063.1

Burnham, Morey, Zhao Ma and Baoqing Zhang. 2016. "Making Sense of Climate Change: Hybrid Epistemologies, Socio-natural Assemblages and Smallholder Knowledge." *Area* 48(1): 18–26.

Chambers, Jeanne C. and M. Pellant. 2008. "Climate Change Impacts on Northwestern and Intermountain United States Rangelands." *Rangelands* 30: 29–33.

Eisen, Michael B. and Patrick O. Brown. 2022. "Rapid Global Phaseout of Animal Agriculture Has the Potential to Stabilize Greenhouse Gas Levels for 30 Years and Offset 68 Percent of CO_2 Emissions This Century." *PLoS Climate* 1(2): e0000010.

Fernández-Giménez, María E., Ahmed El Aich, Oussama El Aouni, Ilhame Adrane and Soufiane El Aayadi. 2021. "Ilemchane Transhumant Pastoralists' Traditional Ecological Knowledge and Adaptive Strategies: Continuity and Change in Morocco's High Atlas Mountains." *Mountain Research and Development* 41(4). https://doi.org/10.1659/MRD-JOURNAL-D-21-00028.1

Gerber, Pierre J., Anne Mottet, Carolyn I. Opio, Alessandra Falcucci and Félix Teillard. 2015. "Environmental Impacts of Beef Production: Review of Challenges and Perspectives for Durability." *Meat Science* 109: 2–12. https://doi.org/10.1016/j.meatsci.2015.05.013

Godde, C. M., R. B. Boone, A. J. Ash, K. Waha, L. L. Sloat, P. K. Thornton and M. Herrero. 2020. "Global Rangeland Production Systems and Livelihoods at Threat Under Climate Change and Variability." *Environmental Research Letters* 15(4): 044021. https://doi.org/10.1088/1748-9326/ab7395

Gosnell, Hannah, Kerry Grimm and Bruce E. Goldstein. 2020. "A Half Century of Holistic Management: What Does the Evidence Reveal?" *Agriculture and Human Values* 37(3): 849–867. https://doi.org/10.1007/s10460-020-10016-w

Greenwood, Paul L. 2021. "Review: An Overview of Beef Production from Pasture and Feedlot Globally, as Demand for Beef and the Need for Sustainable Practices Increase." *Animal* 15: 100295. https://doi.org/10.1016/j.animal.2021.100295

Horowitz, Leah S. 2015. "Local Environmental Knowledge." in *The Routledge Handbook of Political Ecology*, 235–248. Routledge.

Huntsinger, Lynn, Larry C. Forero and Adriana Sulak. 2010. "Transhumance and Pastoralist Resilience in the Western United States." *Pastoralism: Research, Policy, and Practice* 1(1): 1–15.

Ingold, Tim. 2000. *The Perception of the Environment: Essays on Livelihood, Dwelling and Skill*. London, UK: Psychology Press.

Ingold, Tim and Terhi Kurttila. 2000. "Perceiving the Environment in Finnish Lapland." *Body & Society* 6(3–4):183–96.

Maher, Anna T., Nicolas E. Quintana Ashwell, Kristie A. Maczko, David T. Taylor, John A. Tanaka and Matt C. Reeves. 2021. "An Economic Valuation of Federal and Private Grazing Land Ecosystem Services Supported by Beef Cattle Ranching in the United States." *Translational Animal Science* 5(3). https://doi.org/10.1093/tas/txab054

Martin, Jeff. 2021. "Between Scylla and Charybdis: Environmental Governance and Illegibility in the American West." *Geoforum* 123: 194–204. https://doi.org/10.1016/J.GEOFORUM.2019.08.015

Nie, Martin and Peter Metcalf. 2016. "National Forest Management: The Contested Use of Collaboration and Litigation." *Environmental Law Reporter News & Analysis* 46: 10208.

Porensky, Lauren M., Barry L. Perryman, Matthew A. Williamson, Matthew D. Madsen and Elizabeth A. Leger. 2018. "Combining Active Restoration and Targeted Grazing to Establish Native Plants and Reduce Fuel Loads in Invaded Ecosystems." *Ecology and Evolution* 8(24): 12533–12546. https://doi.org/10.1002/ece3.4642

Reeves, Matthew Clark, Karen E. Bagne and John Tanaka. 2017. "Potential Climate Change Impacts on Four Biophysical Indicators of Cattle Production from Western US Rangelands." *Rangeland Ecology & Management* 70(5): 529–39. https://doi.org/10.1016/j.rama.2017.02.005

Sala, Osvaldo, Laura Yahdjian, Kris Havstad and Martín R. Aguiar. 2017. "Rangeland Ecosystem Services: Nature's Supply and Humans' Demand." In *Rangeland Systems*, 467–489. Cham: Springer.

Saliman, Aaron and Margiana Petersen-Rockney. 2022. "Rancher Experiences and Perceptions of Climate Change in the Western United States." *Rangeland Ecology & Management* 84: 75–85. https://doi.org/10.1016/j.rama.2022.06.001

Sayre, Nathan F. 2023a. "A History of North American Rangelands." In *Rangeland Wildlife Ecology and Conservation*, edited by L. B. McNew, D. K. Dahlgren and J. L. Beck, 49–73. Cham: Springer International Publishing.

Sayre, Nathan F. 2023b. "Sustaining Rangelands in the 21st Century." *Rangelands* 45(4): 53–59. https://doi.org/10.1016/j.rala.2022.11.001

Swette, Briana and Eric F. Lambin. 2021. "Institutional Changes Drive Land Use Transitions on Rangelands: The Case of Grazing on Public Lands in the American West." *Global Environmental Change* 66: 102220. https://doi.org/10.1016/j.gloenvcha.2020.102220

HOW TO MOBILIZE PUBLIC WILL TO RESOLVE THE CLIMATE CRISIS

Kristin Haltinner and Dilshani Sarathchandra

As sociologists with an interest in the environment, the two of us have been collaborating on climate change research for nearly a decade. Our work primarily focuses on climate change skepticism in the United States, but more recently we have begun to investigate what factors contribute to changing people's minds on the climate crisis. Through doing this work, we have been struck by the shocking absence of meaningful political action on climate change, despite the catastrophic impacts of the crisis globally.

In putting together this volume, we approached some of the top thinkers in social sciences with similar interests and concerns. We asked them not *what* needs to change to resolve the climate crisis (climate scientists have been telling us of the urgent need to reduce carbon emissions since the 1950s) but *how* we can mobilize the public and political will to achieve these much-needed changes. This volume, then, is the result. It is a handbook on how to mobilize to prevent the worst impacts of the climate crisis – a way to showcase what is possible and to call everyone into the fight for the climate. In this concluding chapter, we consider what these scholars offer and what we now know about magnifying popular will to leverage meaningful change.

Step 1: Recognizing and rejecting the toxic stories we have been socialized to accept

The first section of the book explores the toxic cultural stories that we've all been taught. Pellow, Osborn, and Ergas investigate the foundational stories that shape Western society's practices: neoliberalism, settler colonialism, and gender domination.

DOI: 10.4324/9781003437345-25

These interconnected stories originate at a similar point in human history. The 16th and 17th centuries included a cultural shift in Europe – moving toward attempts to classify and categorize all elements of life in the name of "science." Doing so served to justify the widespread exploitation and abuse of people of color worldwide through colonization, slavery, and genocidal practices. This "Scientific Revolution" led to a cultural acceptance emphasizing difference between invented "types" of people and between people and the more-than-human world. Scientists of the time constructed hierarches, placing certain "types" of people on top – men, White people, Christians, those with land or wealth – and others at the bottom – women, people of color, animals, the earth. Such hierarchies served to justify the exploitation of those on the bottom by those on the top. These ideas were subsequently built into an economic system that continued this racist abuse, gender domination, and extractive and exploitative practices. As such, social inequality has emerged alongside environmental destruction.

The reason these ideas maintain their power is because they have become accepted as normal. In social science we talk about things that are "hegemonic," things that are so dominant in our way of thinking that we fail to even see them and, thus, fail to challenge their continued impacts on society. Neoliberalism, settler colonialism, and gender dominance are idea systems so deeply embedded in our culture that most people fail to recognize their operation and influence. They become accepted as natural, normal, and inevitable – though they are anything but.

The authors in the first section of the book explore different aspects of these hidden cultural stories that serve as drivers of the connected "wicked problems" (see Chapter 7) of social inequality and the climate crisis and offer alternative frameworks for effective organizing. They demonstrate how organizing must be rooted in

1) true democracy, ensuring social equity and justice, and taking on the fossil fuel industry head on (see Chapter 1);
2) becoming earthbound – reengaging with our shared connections to the earth and to one another (see Chapter 2); and
3) educating and empowering women and focusing on a culture of care (see Chapter 3).

Step 2: Recognizing existing use of alternative stories

As mentioned earlier, the success of persistent toxic ideologies is rooted in the way they are rendered invisible to most people in society. There are, always, people who are able to recognize and challenge dominant systems of ideas. Indeed, social change happens when voices at the margins become mainstream, when cultural ideas about what is true, natural, and normal are questioned and replaced with new values (Ahmed 2021). Mobilizing people to climate action requires a new set of cultural stories, one already present in climate organizing on the margins. Effective efforts will

1) emphasize our shared role in environmental action, as exemplified by the dugnad tradition in Norway (see Chapter 4);
2) engage and amplify the power of young activists (see Chapter 5); and
3) ensure a queering of climate change, centering the successful work of queer activism and the need for a queer climate justice (see Chapter 6).

Step 3: Changing the stories

Social theorist Sara Ahmed (2021) examines how cultural change happens. She contends that cultural stories change as new ideas arise from the margin and push their way through fissures and weaknesses in old ways of thinking. The third section of our book considered the question of *how* we amplify stories on the margins to change our cultural narratives. Noll, Haltinner, Sarathchandra, and DeWaard offer insights from social sciences on how cultural change happens and the best ways to engage people in organizing. From these scholars, we learned that such change requires

1) calling climate change what it is, a "wicked problem," emphasizing shared responsibility to motivate action (see Chapter 7);
2) rejecting neoliberal stories of individual responsibility but recognizing that collective action occurs through the united efforts of individuals (see Chapter 8);
3) considering the way that messages are shared about climate change – de-emphasizing paralyzing emotions such as fear while promoting motivating ones like agency, empathy, and hope, and capitalizing on pro-environmental identities to inspire action (see Chapter 9); and
4) employing social-psychological theories to motivate engagement – accepting that which is and committing to change in words, actions, and values (see Chapter 10).

Step 4: Amplifying stories on the margins

Given the urgency of the climate crisis, part of climate organizing needs to be hurrying these changes and amplifying the stories on the margins. In this section, our contributors offer tactics we can use to expedite social change. Though this list is not exhaustive, it offers ideas to begin such efforts. These examples include

1) recognizing the ways that settler colonialism has served to justify a culture of exploitation and extraction and silence Indigenous knowledge and, instead, amplifying tribal voices and knowledge (see Chapter 11);

2) changing international legal conditions that enable environmental destruction by holding those who contribute to ecological destruction responsible via a restorative justice model (see Chapter 12); and

3) engaging media to promote new stories (see Chapter 13).

Step 5: Organizing through a new ethic

Cultural change can be glacially slow, but that's no excuse for inaction. Rather, climate mobilizing needs to put our new ethics – of community care, democracy, equity, and earthboundness – at the center. We need to reject what has been and create and live the reality we seek. To that end, the final section of this book presents ways in which this can be done in practice, through

1) rooting climate activism in Indigenous knowledge and practices (see Chapter 14),

2) connecting with the communities in which novel energy systems are to be implemented (see Chapter 15),

3) truly democratizing movements to ensure that an energy revolution is part of a collaborative project of justice (see Chapter 16),

4) building trust and connection with conservative rural communities (see Chapter 17), and

5) listening to localized knowledge of communities (see Chapter 18).

In short, mobilizing to prevent the worst impacts of the climate crisis must take into account the toxic and dysfunctional cultural stories to which most people adhere; listen to and amplify the voices of those with alternative ethics; promote these stories in line with ways we know motivate human behavior; and center these values of democracy, equity, trust, and care in all efforts of collective action. It is through these processes that cultural narratives will change from ones focused on exploitation and dominance to those rooted in collectivity, connection, and love. It is essential that we start putting these ethics at the forefront of organizing in the present and do the work to create the meaningful cultural change necessary to prevent the worst impacts of the climate crisis.

References

Ahmed, Sara. 2021. "Feminists at Work." *Feminist Killjoys*. Accessed February 20, 2021. https://feministkilljoys.com/2020/01/10/feminists-at-work/

INDEX

Printed in the United States
by Baker & Taylor Publisher Services